Numerical Simulation of Groundwater Flow for the Yakima River Basin Aquifer System, Washington

By D.M. Ely, M.P. Bachmann, and J.J. Vaccaro

Prepared in cooperation with the Bureau of Reclamation, Washington State Department of Ecology, and the Yakama Nation

Scientific Investigations Report 2011–5155

U.S. Department of the Interior
U.S. Geological Survey

U.S. Department of the Interior
KEN SALAZAR, Secretary

U.S. Geological Survey
Marcia K. McNutt, Director

U.S. Geological Survey, Reston, Virginia: 2011

For more information on the USGS—the Federal source for science about the Earth, its natural and living resources, natural hazards, and the environment, visit http://www.usgs.gov or call 1–888–ASK–USGS.

For an overview of USGS information products, including maps, imagery, and publications, visit http://www.usgs.gov/pubprod

To order this and other USGS information products, visit http://store.usgs.gov

Contents

Contents—Continued

Figures

Figures—Continued

Figures—Continued

Tables

Conversion Factors, Datums, and Abbreviations and Acronyms

Conversion Factors

Inch/Pound to SI

Multiply	By	To obtain
Length		
foot (ft)	0.3048	meter (m)
mile (mi)	1.609	kilometer (km)
Area		
acre	4,047	square meter (m^2)
acre	0.004047	square kilometer (km^2)
square mile (mi^2)	2.590	square kilometer (km^2)
Volume		
acre-foot (acre-ft)	1,233	cubic meter (m^3)
Flow rate		
acre-foot per year (acre-ft/yr)	1,233	cubic meter per year (m^3/yr)
foot per day (ft/d)	0.3048	meter per day (m/d)
square foot per day	0.3048	square meter per day
cubic foot per second (ft^3/s)	0.02832	cubic meter per second (m^3/s)
cubic foot per second per square mile [(ft^3/s)/mi^2]	0.01093	cubic meter per second per square kilometer [(m^3/s)/km^2]
inch per year (in/yr)	25.4	millimeter per year (mm/yr)
Hydraulic conductivity		
foot per day (ft/d)	0.3048	meter per day (m/d)
Hydraulic gradient		
foot per mile (ft/mi)	0.1894	meter per kilometer (m/km)
Transmissivity*		
foot squared per day (ft^2/d)	0.09290	meter squared per day (m^2/d)

Temperature in degrees Celsius (°C) may be converted to degrees Fahrenheit (°F) as follows:

$$°F=(1.8×°C)+32$$

Vertical coordinate information is referenced to North American Vertical Datum of 1988 (NAVD 88).

Horizontal coordinate information is referenced to the North American Datum of 1983 (NAD 83).

Datums

Altitude, as used in this report, refers to distance above the vertical datum.

Transmissivity: The standard unit for transmissivity is cubic foot per day per square foot times foot of aquifer thickness [(ft^3/d)/ft^2]ft. In this report, the mathematically reduced form, foot squared per day (ft^2/d), is used for convenience.

Abbreviations and Acronyms

CHD	constant (specified) head cell
CPMod	Columbia Plateau model
CRBG	Columbia River Basalt Group
DEM	digital elevation model
ECM	existing-conditions model
HGU	hydrogeologic unit
HUF	Hydrogeologic-Unit Flow
Kh	horizontal hydraulic conductivity
$K_h{:}K_v$	ratio of horizontal hydraulic conductivity to vertical hydraulic conductivity
K_v	vertical hydraulic conductivity
NWIS	National Water Information System
PRMS	Precipitation-Runoff Modeling System
Reclamation	Bureau of Reclamation
SFR	streamflow-routing cell
TWSA	total water supply available
USGS	U.S. Geological Survey
WaDOE	Washington State Department of Ecology

Numerical Simulation of Groundwater Flow for the Yakima River Basin Aquifer System, Washington

By D.M. Ely, M.P. Bachmann, and J.J. Vaccaro

Abstract

A regional, three-dimensional, transient numerical model of groundwater flow was constructed for the Yakima River basin aquifer system to better understand the groundwater-flow system and its relation to surface-water resources. The model described in this report can be used as a tool by water-management agencies and other stakeholders to quantitatively evaluate proposed alternative management strategies that consider the interrelation between groundwater availability and surface-water resources.

The model was constructed using the U.S. Geological Survey finite-difference model MODFLOW. The model uses 1,000-foot grid cells that subdivide the model domain by 600 rows and 600 columns. Forty-eight hydrogeologic units in the model are included in 24 model layers. The Yakima River, all major tributaries, and major agricultural drains are included in the model as either drain cells or streamflow-routing cells. Recharge was estimated from previous work using physical process models. Groundwater pumpage specified in the model is derived from monthly pumpage values previously estimated from another component of this study. The pumpage values include estimates for wells with standby/reserve rights that are used in drought years.

The model was calibrated to the transient conditions for October 1959 to September 2001. Calibration was completed by using traditional trial-and-error methods and automated parameter-estimation techniques. The model simulates the shape and slope of the water table that generally is consistent with mapped water levels. At well observation points, the average difference between simulated and measured hydraulic heads is -49 feet with a root-mean-square error divided by the total difference in water levels of 4 percent. Simulated river streamflow was compared to measured streamflow at seven sites. Annual differences between measured and simulated streamflow for the sites ranged from 1 to 9 percent. Calibrated model output includes a 42-year estimate of a monthly water budget for the aquifer system.

Five applications (scenarios) of the model were completed to obtain a better understanding of the relation between pumpage and surface-water resources and groundwater levels. For the first three scenarios, the calibrated transient model was used to simulate conditions without: (1) pumpage from all hydrogeologic units, (2) pumpage from basalt hydrogeologic units, and (3) exempt-well pumpage. The simulation results indicated potential streamflow capture by the existing pumpage from 1960 through 2001. The quantity of streamflow capture generally was inversely related to the total quantity of pumpage eliminated in the model scenarios. For the fourth scenario, the model simulated 1994 through 2001 under existing conditions with additional pumpage estimated for pending groundwater applications. The differences between the calibrated model streamflow and this scenario indicated additional decreases in streamflow of 91 cubic feet per second in the model domain. Existing conditions representing 1994 through 2001 were projected through 2025 for the fifth scenario and indicated additional streamflow decreases of 38 cubic feet per second and groundwater-level declines.

Introduction

Surface water in the Yakima River basin, in south-central Washington (fig. 1) is under adjudication. The amount of surface water available for appropriation is unknown, but there are increasing demands for water for municipal, fisheries, agricultural, industrial, and recreational uses. These demands must be met by groundwater withdrawals and/or by changes in the way water resources are allocated and used. On-going activities in the basin for enhancement of fisheries and obtaining additional water for agriculture may be affected by groundwater withdrawals and by rules implemented under the Endangered Species Act for salmonids that have been either listed or were proposed for listing in the late 1990s. An integrated understanding of the groundwater-flow system and its relation to the surface-water resources is needed in order to implement most water-resources management strategies in the basin. In order to gain this understanding, a study of the Yakima River basin aquifer system began in June 2000. The study is a cooperative effort of the U.S. Geological Survey (USGS), Bureau of Reclamation (Reclamation), the Yakama Nation, and the Washington State Department of Ecology (WaDOE).

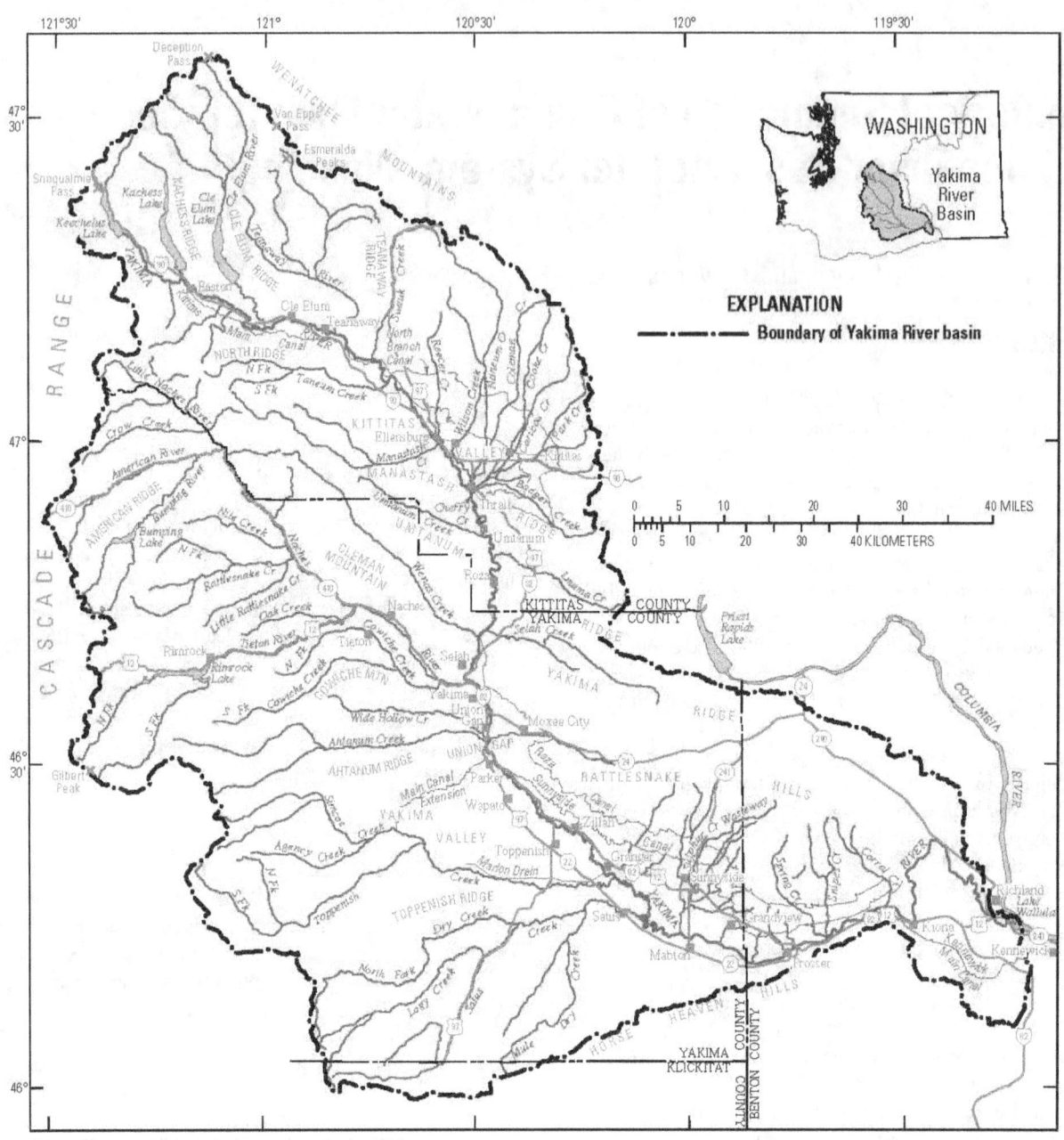

Base modified from U.S. Geological Survey data and other digital sources
Universal Transverse Mercator projection, Zone 10
Horizontal Datum: North American Datum of 1927 (NAD 27)

Figure 1.   The Yakima River basin, Washington.

The overall objectives of the study are to provide a comprehensive description of the groundwater-flow system and its interaction with and relation to surface water, and to provide baseline information for a management tool—a numerical model of the system. The conceptual model of the flow system and the results of the study can be used to guide and support actions taken by management agencies with respect to groundwater availability and to provide information to other stakeholders and interested parties. The numerical model is an integrated tool that can be used to assess short-term to long-term management activities, including the testing of the potential effects of alternative management strategies for water development and use.

The study includes three phases. The first phase includes (1) project planning and coordination, (2) compiling, documenting, and assessing available data, and (3) initial data collection. The second phase consists of data collection to support the following Phase 2 work elements: (1) mapping of hydrogeologic units, (2) estimating groundwater pumpage, (3) developing estimates of groundwater recharge, (4) assessing groundwater-surface water interchanges, and (5) constructing maps of groundwater levels. Together, these five elements provide the information needed to describe the groundwater-flow system, develop the conceptual model, and provide the building blocks for the hydrogeologic framework. In the third phase, a regional-scale numerical model of the groundwater-flow system was constructed. The numerical model was used to enhance understanding of the flow system (including a water budget for the aquifer system) and its relation to surface water, and to test alternative management strategies.

The results of selected work elements of this study have been described in a series of reports (Jones and others, 2006; Vaccaro and Sumioka, 2006; Vaccaro and Maloy, 2006; Vaccaro, 2007; Vaccaro and Olsen, 2007a,b; Jones and Vaccaro, 2008; Vaccaro and others, 2008; Keys and others, 2008; Vaccaro and others, 2009).

Purpose and Scope

This report describes the construction, calibration, and application of the regional numerical model of the groundwater-flow system. The framework for constructing the model was documented in the series of reports listed above. The purpose for constructing the model was to provide an improved understanding of the groundwater-flow system and its relation to surface-water resources. The model development is presented and described, and includes information on the spatial and temporal discretization of the aquifer system, boundary conditions, stresses, and hydraulic properties of the hydrogeologic units constituting the aquifer system. Prior and existing conditions for the period October 1959 to September 2001 were simulated to provide a better understanding of past and current demands on groundwater and surface-water resources in the study area. This existing conditions model (ECM) was then used to simulate potential future impacts given additional demands on groundwater and the potential effects of these simulated increased demands on surface-water resources. Model-calibration methods and results, model limitations, and a sensitivity analysis are documented, as are the observations used to calibrate the model. A regional budget of the groundwater system is documented based on model simulations. The results from the application of the model for five scenarios are also presented, and the results are described relative to differences in simulated streamflow from the calibrated model.

Description of Study Area

The location and setting of the Yakima River basin, a summary of the development of water resources in the basin, and an overview of the geology, based on work by Vaccaro and others (2009) are presented to provide a general background for understanding the study area.

Location and Setting

The Yakima River basin encompasses about 6,200 mi^2 in south-central Washington (fig. 1). The Yakima River basin produces a mean annual unregulated streamflow (adjusted for regulation and without diversions or returns) of about 5,600 ft^3/s [4.1 million acre-ft or about 0.9 (ft^3/s)/mi^2] and a regulated streamflow of about 3,600 ft^3/s [2.6 million acre-ft or about 0.6 (ft^3/s)/mi^2]. The basin includes three Washington State Water Resource Inventory Areas (WRIAs 37, 38, and 39), part of the Yakama Nation lands, and three ecoregions (Cascades, Eastern Cascades, and Columbia Basin—Omernik, 1987; Cuffney and others, 1997). Almost all of Yakima County, more than 80 percent of Kittitas County, and about 50 percent of Benton County are in the basin. Less than 1 percent of the basin, principally in an unpopulated upland area, lies in Klickitat County.

The headwaters of the basin are on the upper, humid east slope of the Cascade Range, where mean annual precipitation is more than 120 in. The basin terminates at the confluence of the Yakima and Columbia Rivers in the low-lying, arid part of the basin that receives about 6 in. of precipitation per year.

Altitudes in the basin range from 400 to nearly 8,000 ft. Eight major rivers and numerous smaller streams are tributary to the Yakima River (fig. 1)—the largest of which is the Naches River. Most of the precipitation in the basin falls during the winter months as snow in the mountains. The mean annual precipitation over the entire basin is about 27 in. (about 12,300 ft³/s or 8.9 million acre-ft). The spatial pattern of mean annual precipitation resembles the pattern of the basin's highly variable topography. The difference between the mean annual precipitation and mean annual unregulated streamflow is 6,700 ft³/s (about 4.8 million acre-ft). On the basis of this difference and the simplifying assumptions of only small net groundwater inflow to or outflow from the basin and negligible mean annual changes in groundwater storage within the basin, about 54 percent of the precipitation is consumed by evapotranspiration under natural conditions.

The basin is separated into several broad valleys by east-west trending anticlinal ridges. The valley floors slope gently towards the Yakima River. Few perennial tributary streams traverse these valleys. Most of the population and economic activity occurs in these valleys.

Irrigated agriculture is the principal economic activity in the basin. The average annual surface-water demand met by Reclamation's Yakima Project is about 2.5 million acre-ft, and there is an additional 336,000 acre-ft of demand in the lower part of the basin. More than 95 percent of the surface-water demand is for irrigation of about 500,000 acres in the low-lying semiarid to arid parts of the basin (fig. 2). The demand is partly met by storage of nearly 1.1 million acre-ft of water in six Reclamation reservoirs. The major management point for Reclamation is the streamflow gaging station at the Yakima River near Parker at river mile 103.7 (USGS station No. 12505000, fig. 3); this site is just below the Sunnyside and Wapato (main) canal diversions. Just upstream of this site at about river mile 106.8 is the location that is considered the dividing line between the upper (mean annual precipitation of 7–145 in.) and lower (mean annual precipitation of 6–45 in.) parts of the Yakima River basin. About 45 percent of the water diverted for irrigation is eventually returned to the river system as surface-water inflows and groundwater discharge, but at varying time-lags (Bureau of Reclamation, 1999). During the low-flow period, these return flows, on average, account for about 75 percent of the streamflow below the streamflow-gaging station near Parker. Most of the surface-water demand in the basin below Parker is met by these return flows and not by the release of water from the reservoirs. As a result of water use in the basin, the difference between mean annual unregulated (5,600 ft³/s) and regulated (3,600 ft³/s) streamflow in the basin is about 2,000 ft³/s, suggesting that some 1.4 million acre-ft of water, or about 16 percent of the precipitation in the basin, is consumptively used—principally by irrigated crops through evapotranspiration.

Development of Water Resources

Missionaries arrived in the basin in 1848 and established a mission in 1852 on Atanum (now Ahtanum) Creek. They were some of the first non-Indian settlers to use irrigation on a small scale. Miners and cattlemen immigrated to the basin in the 1850s and 1860s, which resulted in a new demand for water. With increased settlement in the mid-1860s, irrigation of the fertile valley bottoms began and the outlying areas were extensively used for raising stock. One of the first known non-Indian irrigation ditches was constructed in 1867 and diverted water from the Naches River (Parker and Storey, 1916; Flaherty, 1975). Private companies later delivered water through canal systems built between 1880 and 1904 for the irrigation of large areas. The development of irrigated agriculture was made more attractive by the construction of the Northern Pacific Railway that reached Yakima in December 1884 and provided a means to transport agricultural goods to markets; two years later, the completion of the railway to the Pacific coast provided new and easily accessible markets for agricultural products. The State of Washington was created in 1889, spurring further growth in the basin, especially because the cities of Ellensburg and Yakima were in contention for being the state capital. By 1902, about 120,000 acres were under irrigation, mostly by surface water (Parker and Storey, 1916; Bureau of Reclamation, 1999).

The Federal Reclamation Act of 1902 enabled the construction of Federal water projects in the western United States in order to expand the development of the West. In 1905, the Washington State Legislature passed the Reclamation Enabling Act, and the Yakima Federal Reclamation Project was authorized to construct facilities to irrigate about 500,000 acres. As part of the 1905 authorization and extensions, all forms of further appropriation of unappropriated water in the basin were withdrawn (Parker and Storey, 1916). Six dams were constructed as part of the Yakima Project: Bumping Dam in 1910, Kachess Dam in 1912, Clear Creek Dam in 1914, Keechelus Dam in 1917, Tieton Dam (Rimrock Lake) in 1925, and Cle Elum Dam in 1933. The six reservoirs have a total capacity of about 1.07 million acre-ft, of which about 78 percent is stored in the upper arm of the Yakima River and 22 percent is stored in the Naches River arm. The construction of the dams and other irrigation facilities resulted in an extremely complicated surface-water system (fig. 3). These Federal reservoirs provide water storage to meet irrigation requirements of the major irrigation districts during the period when the natural streamflow from unregulated streams can no longer meet demands; the onset of this period is referred to as the 'storage control' date. Several of the reservoirs also provide instream flows during the winter for the incubation of salmon eggs in redds (gravel spawning nests).

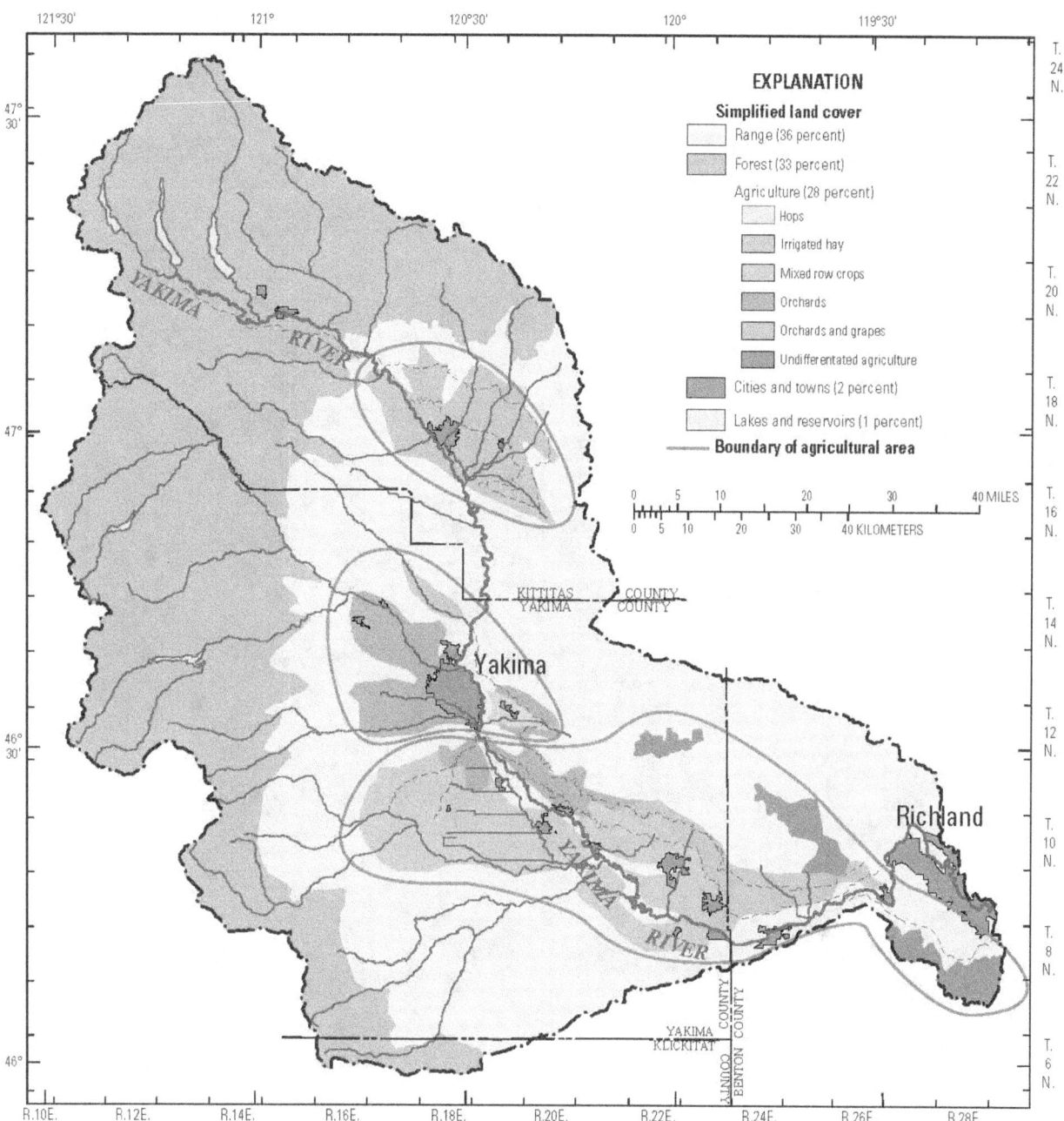

Figure 2.   Land use and land cover, Yakima River basin, Washington, 1999.

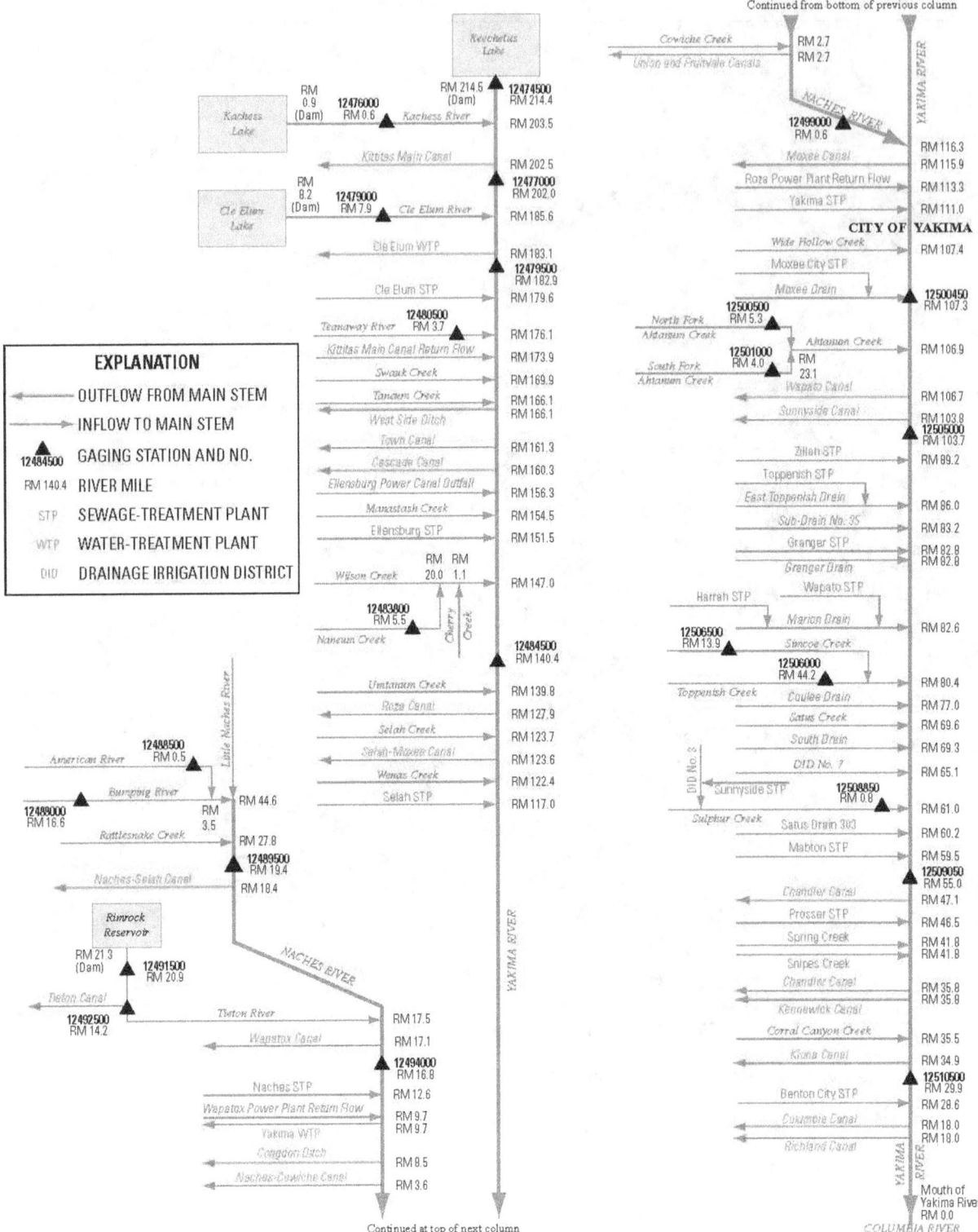

Figure 3.   Selected tributaries, diversion canals, return flows, and stream-gaging stations, Yakima River basin, Washington.

Legal challenges to water rights resulted in the 1945 Consent Decree (U.S. District Court, 1945) that established the framework of how Reclamation operates the Yakima Project to meet water demands. The Decree determined three classes of rights—nonproratable (priority dates of pre-May 1905), proratable (priority date of May 1905 when Reclamation obtained the unappropriated water), and junior (post-May 1905 priority dates). When the total water supply available (TWSA, defined as current available storage in the reservoirs, forecasted estimates of unregulated flow, and other sources that are principally return flows) is not sufficient to meet all classes of rights, the proratable rights are decreased according to the quantity of water available estimated by the TWSA, and junior users can be completely turned off. As of 2008, the years when proration levels were defined were 1973, 1977, 1979, 1987–88, 1992–94, 2001, and 2005. This legally mandated method, which was upheld in a 1990 court ruling, generally performs well in most years, but is dependent on the accuracy of the TWSA estimate. In some years, for example 1977, problems have arisen because of errors in the TWSA estimate (Kratz, 1978; Glantz, 1982). In addition, numerous proratable users have obtained groundwater-water rights to pump supplemental water in the years that they receive prorated quantities of surface water. System management also accounts for defined instream flows at selected target points on the river, and for suggested changes in storage releases recommended by the Systems Operations Advisory Committee (SOAC)—the advisory board of fishery biologists representing the different stakeholders (Systems Operations Advisory Committee, 1999). The operations for meeting instream flows are most affected by a 1980 Federal circuit court decision and by Title XII of a Public Law that instituted (beginning in 1995) new instream flows for the former and target flows for the latter at two diversions dams (Sunnyside and Prosser). The 1980 decision resulted in lower reservoir releases from the Keechelus and Cle Elum reservoirs in mid-September to prevent spawning chinook from building redds at higher altitudes along the channel. To meet demands after mid-September, releases from the Naches arm reservoirs are increased. This operational procedure is called 'flip-flop'.

The drilling of numerous wells for irrigation was spurred by new (post 1945) well-drilling technologies, legal rulings, and the onset of a multi-year dry period in 1977 (Vaccaro, 1995, 2002). Population growth in the basin remains the driving force behind the increased drilling of shallow domestic wells and deeper public water supply wells. Vaccaro and Sumioka (2006) estimated that on the order of 45,000 water wells currently are in the basin of which information for about 20,000 have been compiled (fig. 4).

About 70 percent of these wells are shallow (10–250 ft deep). On the basis of the digital water-rights database provided by WaDOE (R. Dixon, Washington State Department of Ecology, written commun., 2001) and other information, 2,874 active groundwater rights are associated with the wells in the basin that can legally collectively withdraw an annual quantity of about 529,231 acre-ft during dry years. This quantity is the legal paper right but not the actual withdrawal quantity. The irrigation rights are for the irrigation of about 129,570 acres. There are about 16,600 groundwater claims in the basin; these claims are for some 270,000 acre-ft of groundwater (J. Kirk, Washington State Department of Ecology, written commun., 1998). 'A water right claim is a statement of claim to water use that began before the state Water Codes were adopted, and is not covered by a water right permit or certificate. A water right claim does not establish a water right, but only provides documentation of one if it legally exists. Ultimately, the validity of claimed water rights would be determined through general water right adjudications' (Washington State Department of Ecology, 1998). A groundwater claim means a user claims that they were using groundwater continuously for a particular use, prior to 1945, when the State legislature enacted the Ground Water Code.

Overview of the Geology

The Columbia Plateau has been informally divided into three physiographic subprovinces (Meyers and Price, 1979). The western margin of the Columbia Plateau contains the Yakima Fold Belt subprovince and includes the Yakima River basin. The Yakima Fold Belt is a highly folded and faulted region, and within the study area it is underlain by various consolidated rocks ranging in age from Precambrian to Tertiary, and unconsolidated materials and volcanic rocks of Quaternary age. The simplified surficial geology (Fuhrer and others, 1994) clearly shows the wide variety of rock materials present in the basin (fig. 5). In the Yakima River basin, the headwater areas in the Cascade Range include metamorphic, sedimentary, and intrusive and extrusive igneous rocks. The central, eastern, and southwestern parts of the basin comprise basalt lava flows of the Columbia River Basalt Group (CRBG) with some intercalated sediments that are discontinuous and weakly consolidated. The lowlands are underlain by unconsolidated and weakly consolidated valley-fill comprising glacial, glacio-fluvial, lacustrine, and alluvium deposits that in places exceed 1,000 ft in thickness (Drost and others, 1990). Wind-blown deposits, called loess, are present locally along the lower valley.

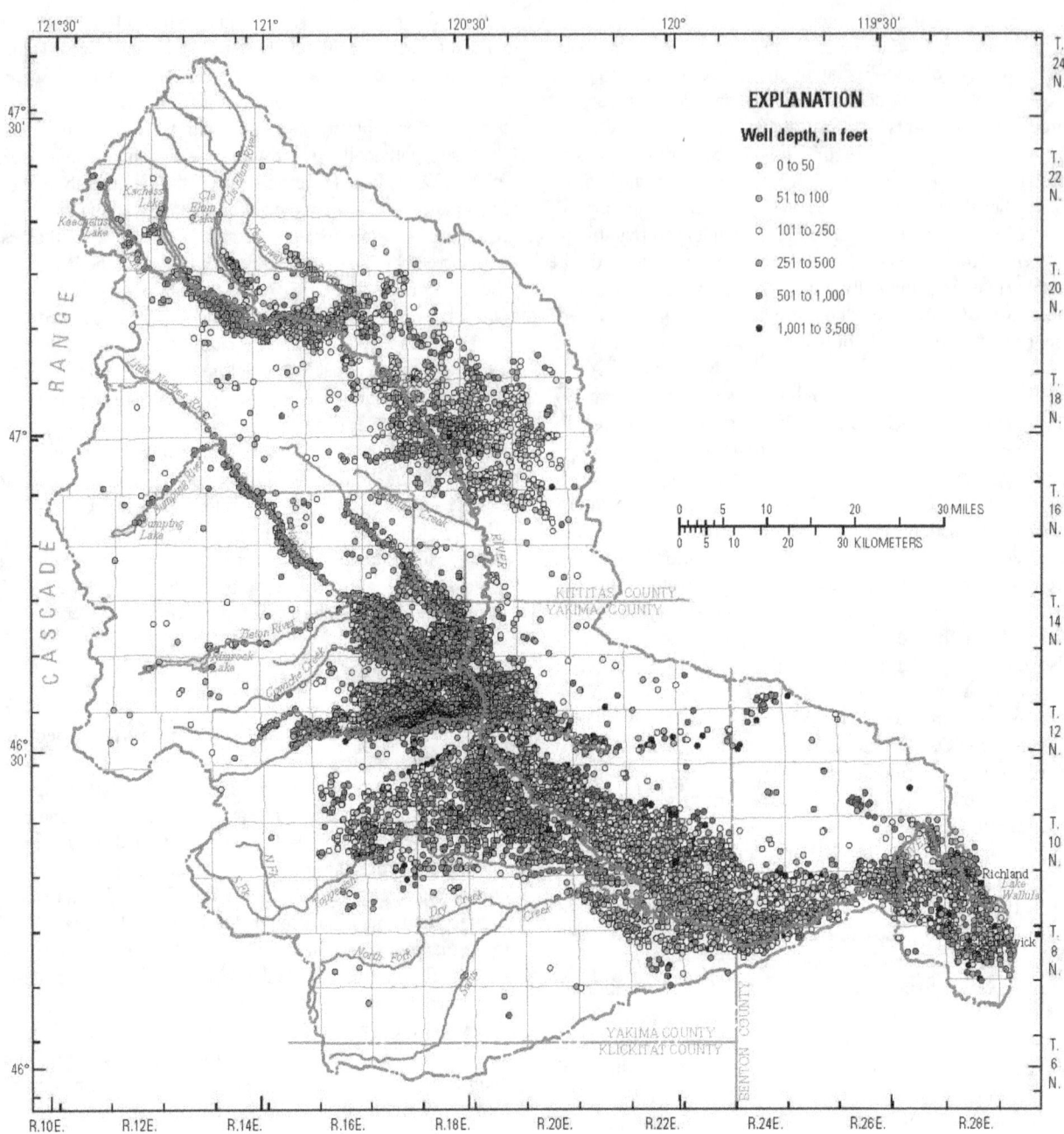

Figure 4.   Distribution of depths of water wells, Yakima River basin, Washington.

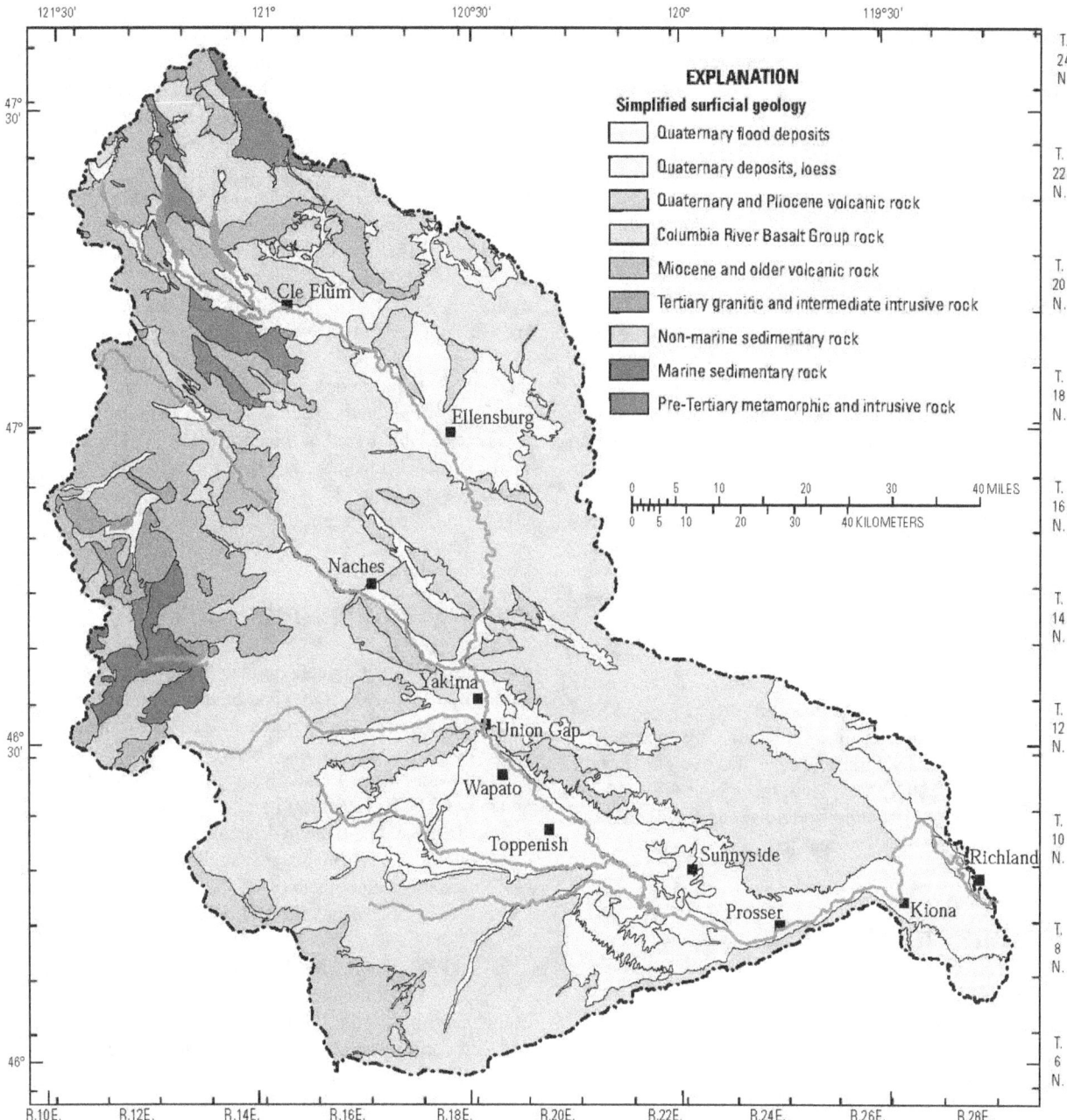

Figure 5. Simplified surficial geology, Yakima River basin, Washington.

Valley-fill deposits and basalt lava flows are important for groundwater occurrence in the study area. The basalt comprises a series of flows erupted during various stages of the Miocene Age, from 17 to 6 million years ago. Basalt erupted from fissures located in the eastern part of the Columbia Plateau and individual flows range in thickness from a few feet to more than 100 ft. The total thickness in the central part of the plateau is estimated to be greater than 10,000 ft (Drost and others, 1990) and the maximum thickness in the Yakima River basin is more than 8,000 ft. Unlike most of the Columbia Plateau, the CRBG in the Yakima Fold Belt is underlain by sedimentary rocks. The valley-fill deposits were eroded from the Cascade Range and from the east-west-trending anticlinal ridges that were formed by the buckling of the basalt sequence during mid- to late-Miocene time. Most of these deposits are part of the Ellensburg Formation. This formation underlies, intercalates, and overlies the basalts along the western edge, and constitutes most of the thickness of the unconsolidated deposits (informally called the overburden; Drost and others, 1990) in the basins. The basins are narrow to large open synclinal valleys between the numerous anticlinal ridges.

The deposition of a thick, upper sequence of sand, gravel, and some fine-grained material is the result of erosion by glacial ice and transport by meltwater streams. Damming of large lakes by glacial ice during the Pleistocene epoch resulted in the deposition of silt and clay beds in parts of the uplands. When the lakes drained, the fine sediments were exposed, subsequently eroded by wind, and deposited over the lower, eastern parts of the study area. Thus, the unconsolidated materials in the basins abutting and interbedded with the basalts range in age from Miocene to Holocene.

Well Data and Their Use

Four sets of mass water-level measurements in wells were made in autumn 2000, spring 2001, autumn 2001, and spring 2002. In addition, Yakama Nation staff made monthly measurements at numerous wells, as well as weekly measurements at nine wells during June 2000–June 2002. The measured wells were selected to obtain a good spatial distribution, both laterally and vertically; the location of wells with water-level measurements during the period of interest is shown in figure 6. In addition, historical water-level measurements made by WaDOE and Yakama Nation that had not been entered into the National Water Information System (NWIS) were compiled. Information and water levels for all these wells were put into NWIS.

Additional historical water levels were available from more than 8,000 wells. A subset of these water levels, which is in NWIS, was measured by the USGS. A second subset of these water levels also is in NWIS, and represents water levels reported on drillers' well logs. A third subset of water levels consists of water levels reported on 'uncoded' well logs. Uncoded well logs were any well logs in USGS paper files that were not in NWIS. Selected information from more than 9,000 of these uncoded wells was put into digital form in order to provide additional information for: (1) mapping the spatial distribution of wells in the basin (fig. 4), (2) evaluating the spatial distribution of well depths, (3) analyzing specific-capacity tests, (4) identifying flowing wells, (5) mapping water levels, and (6) mapping hydrogeologic units; these items are documented in Vaccaro and others (2009). Information on specific-capacity tests and water levels was not available for all of the uncoded wells. The water-level information provided the basic data for calibrating the groundwater-flow model. The water-level data are described in more detail in a later section in this report.

Hydrogeologic Units

Thirty-four hydrogeologic units were mapped and named in this study by Jones and others (2006) and Jones and Vaccaro (2008). Hydrogeologic unit (HGU) information for the 28 major units is summarized in table 1 and a correlation chart showing the relation between generalized geologic units and mapped HGUs is shown in table 2. Information used to help define and map the hydrogeologic units in table 2 is from well logs and published maps of Swanson and Wright (1978), Swanson and others (1979a, 1979b), Tanaka and others (1979), Tabor and others (1982, 1987, 1993), Walsh (1986a, 1986b), Korosec (1987), Phillips and Walsh (1987), Schasse (1987), Gulick and Korosec (1990), Reidel and Fecht (1994a, 1994b), and Schuster (1994a, 1994b, 1994c). Most of the maps in the post-1980 publications were available in digital form from the Washington State Department of Natural Resources (2002). Additional information for the Hanford Site was provided by S.P. Reidel and P.D. Thorne (Battelle Institute, written commun., 2003 and 2005) and for the Toppenish basin by Newell Campbell (unpublished maps, produced for the Yakama Nation, 2001).

Information about the hydrogeologic units defined and mapped in this study is summarized below. A more detailed description of the study methods and hydrogeologic units can be found in Jones and others (2006) and Jones and Vaccaro (2008). The detailed descriptions include maps of thickness of selected units, depths to the top of selected units, and hydrogeologic cross sections.

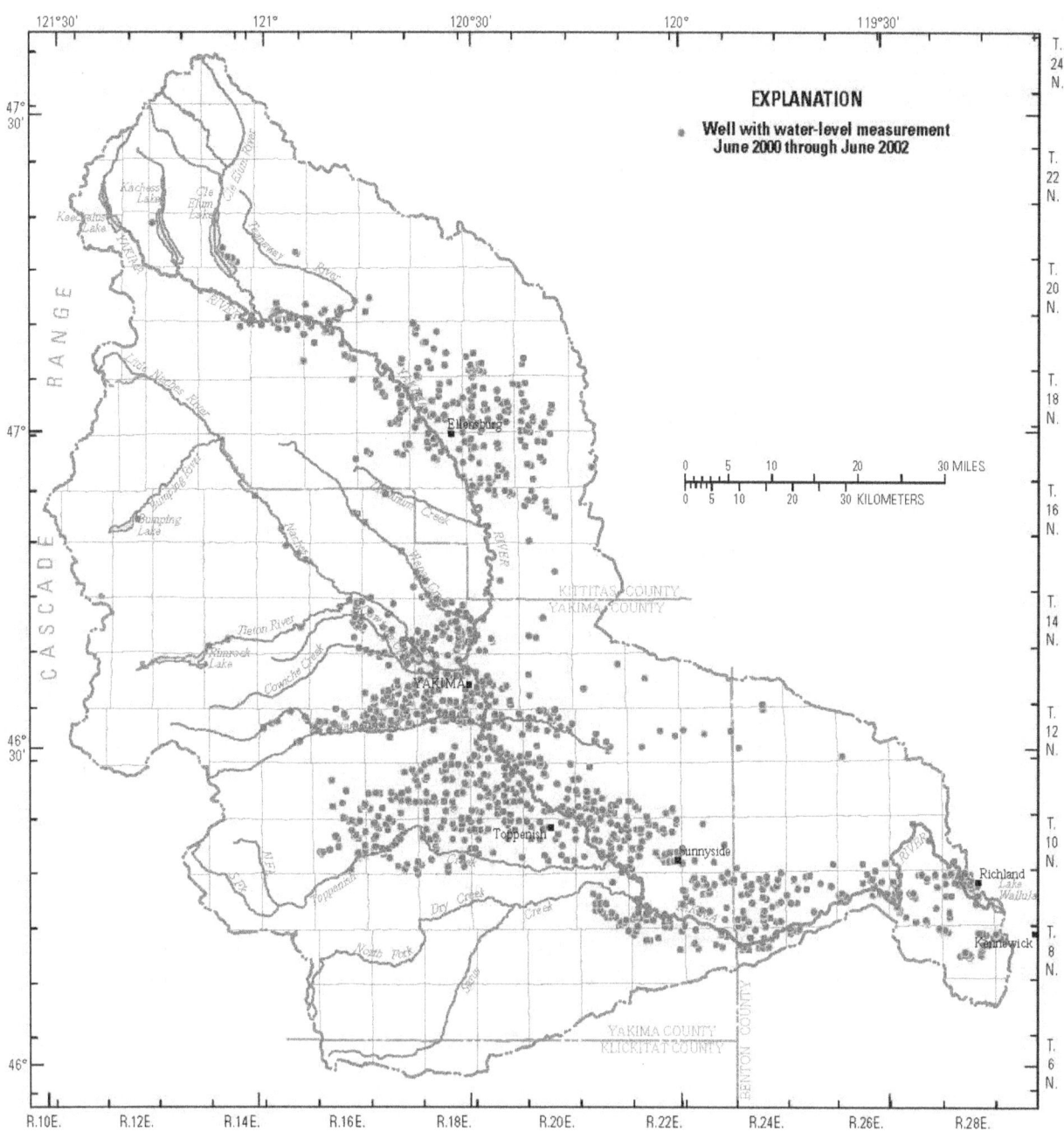

Figure 6.   Location of wells with water-level measurements, Autumn 2000 through Spring 2002, Yakima River basin, Washington.

Table 1.   Information for the hydrogeologic units of the Yakima River basin aquifer system, Washington.

[**Abbreviations:** mi², square mile; ft, foot; –, not calculated]

Structural basin name	Mapped area (mi²)	Hydrogeologic unit	Lithology	Thickness (ft)		
				Range	Average	Median
Roslyn basin	70	1 (alluvial and coarse-grained unconsolidated)	Alluvial, lacustrine, and glacial deposits	0–360	80	80
		2 (fine-grained unconsolidated)	Fine-grained lacustrine clay and silt deposits	0–530	180	170
		3 (coarse-grained unconsolidated)	Coarse-grained sand and gravel deposits	0–240	60	50
		Total basin thickness	All deposits	0–700	150	110
Kittitas basin	270	1 (alluvial)	Floodplain alluvial deposits	0–100	30	10
		2 (unconsolidated)	Loess, alluvial fan, glacial terrace, and Thorp gravel deposits	0–790	180	150
		3 (consolidated)	Ellensburg Formation and undefined continental sedimentary deposits	0–2,040	600	350
		Total basin thickness	All deposits	0–2,120	500	270
Selah basin	170	1 (alluvial)	Floodplain alluvial deposits	0–90	30	30
		2 (unconsolidated)	Loess, alluvial fan, glacial terrace, and Thorp gravel deposits	0–290	50	40
		3 (consolidated)	Ellensburg Formation and undefined continental sedimentary deposits	0–1,920	320	200
		Total basin thickness	All deposits	0–1,920	300	200
Yakima basin	230	1 (alluvial)	Floodplain alluvial deposits	0–120	20	20
		2 (unconsolidated)	Loess, alluvial fan, glacial terrace, and Thorp gravel deposits	0–350	90	80
		3 (consolidated)	Ellensburg Formation and undefined continental sedimentary deposits	0–1,840	510	450
		Total basin thickness	All deposits	0–1,840	530	410
Toppenish basin	440	1 (fine-grained unconsolidated)	Touchet Beds, terrace, loess, and some alluvial deposits	0–80	10	10
		2 (coarse-grained unconsolidated)	Coarse-grained sand and gravel deposits	0–270	90	80
		3 (consolidated)	Consolidated deposits of the upper Ellensburg Formation and undefined continental sedimentary deposits	0–970	350	320
		4 (fine-grained deposits)	Top of Rattlesnake Ridge unit of the Ellensburg Formation or 'Blue Clay unit'	0–520	170	140
		5 (coarse-grained deposits)	Base of Rattlesnake Ridge unit of the Ellensburg Formation	0–140	20	20
		Total basin thickness	All deposits	0–1,210	550	550
Benton basin	1,020	1 (unconsolidated)	Alluvial, alluvial fan, loess, terrace, dune sand, Touchet Beds, Missoula flood, and Ringold Formation deposits.	0–870	120	70
		2 (consolidated)	Ellensburg Formation and undefined continental sedimentary deposits	0–680	100	60
		Total basin thickness	All deposits	0–870	120	60

Table 1. Information for the hydrogeologic units of the Yakima River basin aquifer system, Washington.—Continued

[**Abbreviations:** mi^2, square mile; ft, foot; –, not calculated]

Unit name	Mapped area (mi^2)	Extent (mi^2)	Lithology	Thickness (ft)		
				Range	Average	Median
Columbia River Basalt Group and interbeds						
Saddle Mountains	2,289	457 mi^2 surface outcrop, 1,804 mi^2 below surface, 28 mi^2 not present in mapped area	Saddle Mountains Basalt flow members and interbeds	0–1,110	550	560
Mabton Interbed	2,206	2,179 mi^2 below surface, 27 mi^2 not present in mapped area		0–250	70	70
Wanapum unit	3,444	659 mi^2 surface outcrop, 2,757 mi^2 below surface, 28 mi^2 not present in mapped area	Wanapum Basalt flow members and interbeds	0–1,180	600	490
Vantage Interbed	3,087	3,047 mi^2 below surface, 40 mi^2 not present in mapped area		0–135	30	20
Grande Ronde unit	5,383	1,547 mi^2 surface outcrop, 3,786 mi^2 below surface, 50 mi^2 not present in mapped area	Grande Ronde Basalt flow members and interbeds	–	–	–
Bedrock units						
Quaternary bedrock		82	Principally volcanics, but with minor amounts of sediments	–	–	–
Tertiary		1,300	Sediments, and volcanic and plutonic rocks	–	–	–
Mesozoic		136	Metamorphic, volcanic, and plutonic rocks	–	–	–
Paleozoic		2	Metamorphic rocks	–	–	–
All bedrock units	1,520			–	–	–

Table 2.   Correlation chart showing regional relation between generalized geologic units and hydrogeologic units in the basin-fill and Columbia River Basalt Group units and bedrock units for the Yakima River basin aquifer system, Washington.

[**Hydrogeologic unit:** B, Benton basin; K, Kittitas basin; R, Roslyn basin; S Selah basin; T, Toppenish basin; Y, Yakima basin. See table 1 for additional information on geologic units. **Abbreviations:** Fm., formation; Mtn, mountain; Cr., creek]

BASIN-FILL AND COLUMBIA RIVER BASALT GROUP UNITS					
ERA	PERIOD	EPOCH		SIMPLIFIED GEOLOGIC UNITS	HYDROGEOLOGIC UNIT
CENOZOIC	Quaternary	Holocene		Alluvium, alpine glaciation, alluvial fan, dune sand, artificial fill, and peat deposits	Unit 1 {R, K, S, Y, T, B}
		Pleistocene		Alluvium, alpine glacial drift, alluvial fan, Palouse Fm., Lakedale Drift, Lookout Mountain Ranch Drift, Hayden Creek Drift, Kittitas Drift, Evans Creek Drift, unknown continental sedimentary deposits, dune sand, Missoula glacial lake deposits	Unit 1 {R, K, S, Y, T, B}, Unit 2 {R, K, S, Y, T}
	Tertiary	Pliocene		Alluvial fan, Ringold Fm., Ellensburg Fm., Dalles Fm., Thorpe Gravel, and unknown continental sedimentary deposits	Unit 1 {B}, Unit 2 {R, K, S, Y, T}, Unit 3 {R, K, S, Y}
		Miocene		Ellensburg Fm., Ringold Fm., Dalles Fm., Snipes Mountain deposits, and unknown continental sedimentary deposits	Unit 2 {B}, Unit 3 {R, K, S, Y, T}, Unit 4 {T}, Unit 5 {T}
			Columbia River Basalt Group	Saddle Mountains Basalt flow members and interbeds	Saddle Mountains unit {SM}
				Mabton Interbed	Mabton unit
				Wanapum Basalt flow members and interbeds	Wanapum unit {WN}
				Vantage Interbed	Vantage unit
				Grande Ronde Basalt flow members and interbeds	Grande Ronde unit {GR}

There are four categories of mapped hydrogeologic units: (1) unconsolidated units composed of Pliocene to Holocene sediments that may or may not be subdivided, (2) semi-consolidated to consolidated units (referred to here as consolidated units) composed of Miocene-Pliocene sediments, (3) Miocene CRBG and interbed units, and (4) Paleozoic to Quaternary bedrock units. The first 2 categories include 19 mapped units and 6 subunits of one mapped unconsolidated unit. These two categories consist of basin-fill deposits occurring in six structural-sedimentary basins (fig. 7), herein called structural basins; the geologic structure delineating the basins is clearly defined by the folds and faults shown in figure 7. The structural basins generally are consistent with the groundwater basins of Kinnison and Sceva (1963). Each unit in a structural basin has been named; for example, a unit in the Kittitas basin that is composed of alluvial deposits is named Unit 1 (Jones and others, 2006, table 2). These names, however, may or may not represent the same type of hydrogeologic unit in different structural basins. For each structural basin, the naming of these units starts at 1 for the uppermost unit and increases with

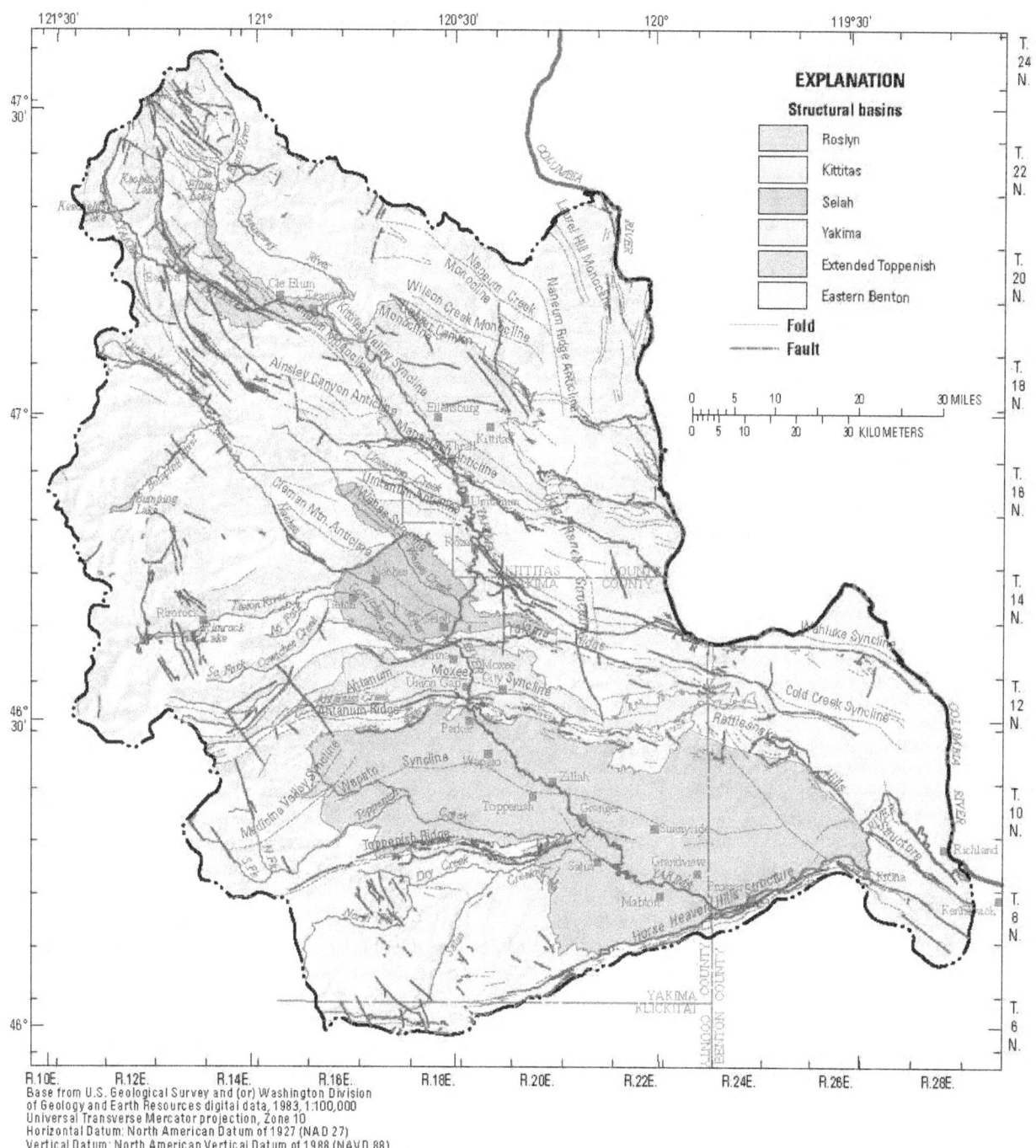

Figure 7. Location of six structural basins and geologic structure, Yakima River basin aquifer system, Washington.

the depth (age) of a unit; thus, if there were three units in a structural basin, they would be numbered 1 through 3 (Unit 1, Unit 2, and Unit 3).

In the Kittitas, Selah, and Yakima basins (fig. 7), Jones and others (2006) mapped an alluvial unit (Unit 1), an unconsolidated unit (Unit 2), and a consolidated unit (Unit 3); the Yakima basin also is locally called the Ahtanum-Moxee basin (Ahtanum and Moxee subbasins). In the Roslyn basin, a surficial coarse-grained aquifer unit (Unit 1), a lacustrine confining unit (Unit 2), and a confined coarse-grained aquifer unit (Unit 3) underlying the confining units were mapped (table 1). In the Toppenish basin, five units were mapped that consist of two units of coarse-grained and fine-grained unconsolidated deposits and three units comprised of part of the Ellensburg Formation. An unconsolidated unit (Unit 1) and a consolidated unit (Unit 2) were mapped in the Benton basin. In addition, in the eastern part of the Benton basin, the unconsolidated unit (the only unit present in this area) was further divided into six subunits (Jones and others, 2006). These six subunits, which occur together only in this part of the study area, consist of fluvial and lacustrine deposits of the Ringold Formation, glaciofluvial sediments of the Hanford Formation, and pre-Missoula Flood gravels. Typically, the consolidated units contain deposits that are more consolidated (for example, sandstone) than the deposits of the unconsolidated units (for example, recent sand and gravel). Thus, a total of 19 units and 6 subunits were mapped for the basin-fill deposits.

The structural-basin boundaries (fig. 7) are slightly different than those shown in Jones and others (2006) in order to provide an improved framework for constructing the groundwater-flow model. These differences include an extension of the Toppenish basin to include the western part of the Benton basin where both unconsolidated and consolidated units are present. The eastern part of the Benton basin containing the six unique subunits also was extended eastward to the Columbia River, to include about 217 mi^2 outside the Yakima River basin. This area generally coincides with the Hanford Site and is part of what is herein called the extended study area and is included in the model domain. The extents of the six eastern Benton basin subunits were obtained from S.P. Reidel and P.D. Thorne (Battelle Institute, written commun., 2003 and 2005). To distinguish these extended areas from the previously defined structural basins, they are referred to as the extended Toppenish basin and the eastern Benton basin (fig. 7).

The unconsolidated units include alluvial, alluvial fan, terrace, glacial, loess, lacustrine, and flood (Touchet Beds) deposits that range from coarse-grained gravels to fine-grained clays, with some cemented gravel (Thorp gravel and similar

unnamed gravels). Most of the unconsolidated units consist of coarse-grained deposits.

The deposits that constitute the consolidated units are principally deposits of the Ellensburg Formation, but also include some undifferentiated continental sedimentary deposits. These units include continental sandstone, shale, siltstone, mudstone, claystone, clay, and lenses or layers of uncemented and weakly to strongly cemented gravel and sand (conglomerate). These clastic deposits are one of the most stratigraphically complex parts of the aquifer system. In the structural basins where these deposits overlie bedrock, the deposits were either mapped as one hydrogeologic unit (called the consolidated unit, for example, Unit 3 in the Kittitas, Selah, and Yakima basins and Unit 2 in the Benton basin) or the deposits were subdivided into several hydrogeologic units as in the Toppenish basin (Jones and others, 2006, table 2). Except for the Mabton and Vantage Interbeds, where the Ellensburg Formation is interbedded with the CRBG, less extensive sedimentary deposits were considered part of the CRBG hydrogeologic units (Jones and Vaccaro, 2008). The Mabton and Vantage Interbeds are considered separate hydrogeologic units and the depth to the top of these units was mapped by Jones and Vaccaro (2008).

The lithology of the consolidated units varies from the Kittitas basin to the extended Toppenish basin; consolidated units are absent in the Roslyn and Eastern Benton basins. The variations are due to spatial-temporal variations in deposition, erosion, and structural deformation (faulting and folding). In some areas, the consolidated unit may consist of principally clay, sand, and gravel. In other areas, the unit may consist of predominantly sandstone with some cemented gravel layers. In the Moxee subbasin of the Yakima basin, a 637-ft well penetrated clay, sandstone, and shale of the consolidated unit to its final depth and only minor amounts of interbedded sands and gravels were present.

The other two categories of units are the Miocene age CRBG units and Paleozoic to Quaternary bedrock units (table 2). The CRBG units form the major, productive volcanic-rock part of the aquifer system. The generally older, non-CRBG bedrock units (herein called bedrock units) are present primarily along the northern, northwestern, and western margins of the basin (fig. 5) and include volcanic, intrusive, marine and nonmarine sedimentary and metamorphic rock units. In a few areas at the western margins of the basin in the High Cascades, younger rock units composed principally of Quaternary volcanics units are present (table 2) that were derived from the volcanically active Cascade Range.

The CRBG contains three formations in the study area, which are, from oldest to youngest, the Grande Ronde, Wanapum, and Saddle Mountains Basalts (table 2). The Grande Ronde Basalt is the most extensive, consisting of 85 to 88 percent of the total volume of the CRBG and the Saddle Mountains Basalt is the least extensive, consisting of less than 2 percent of the total volume of the CRBG (Reidel, 1982; Tolan and others, 1989). The younger Saddle Mountains Basalt contains the thickest and most sedimentary interbeds, especially in the Yakima Fold Belt, because of its episodic eruptions and proximity to source material. For example, a 773-ft deep well penetrated the Saddle Mountains Basalt from 17 to 70 ft and then penetrated clay, sand, sandstone, and gravel to 587 ft, followed by 34 ft of basalt that was underlain by 152 ft of interbeds. A nearby well penetrated 400 ft of sand and clay before penetrating 50 ft of the Saddle Mountains Basalt; 270 ft of sand and clay underlies the basalt to the final depth of 720 ft. There are locations where the interbeds in the Saddle Mountains Basalt constitute more than 50 percent of the total thickness of the formation.

Each CRBG formation was defined as a hydrogeologic unit (table 2). In this report the formations are called units because they include interbeds. Each unit includes a thick sequence of basalt flows that overlap and intermingle. In turn, individual basalt flows comprise the members of the formations. The depth to the top of each unit was mapped by Jones and Vaccaro (2008), and the configuration of the top of each unit is very complex. The Mabton and Vantage Interbeds also were defined as hydrogeologic units and depths to their tops were mapped by Jones and Vaccaro (2008). The Mabton unit lies between the Saddle Mountains and Wanapum units and the Vantage unit separates the Wanapum and Grande Ronde units. Neither the Mabton nor the Vantage units are present throughout the extent of the overlying CRBG units. The maps showing the depth to the tops include the area east of the Yakima River basin boundary to the Columbia River, an area of about 700 mi². This additional area, combined with the Yakima River basin, is called the extended study area and encompasses about 6,900 mi²; this area defines the domain of the groundwater-flow model. The extended study area thus is bounded on the east by the Columbia River, a hydrologic

boundary in the regional groundwater-flow model that is described later in the report.

Within the model domain, the lateral extent of the Grande Ronde unit is about 5,383 mi², the extent of the Wanapum unit is about 3,444 mi² and has an average thickness of about 600 ft, and the extent of the Saddle Mountains unit is about 2,289 mi² and has an average thickness of about 550 ft (table 1).

Four bedrock units were defined on the basis of their age. These units are called the Paleozoic unit, the Mesozoic unit, the Tertiary unit, and the Quaternary bedrock unit. The Paleozoic unit consists of metamorphic rocks and the Mesozoic unit has three subdivisions consisting of metamorphic, volcanic, and plutonic rocks. The Tertiary unit also has three subdivisions consisting of sedimentary, volcanic, and plutonic rocks. The Quaternary bedrock unit consists principally of volcanics, but in a few small areas it consists of sediments. Where not overlain by younger deposits, the surficial extents of these bedrock units were obtained from the digital GIS database for Washington (Washington State Department of Natural Resources, 2002). Neither thickness nor vertical subdivisions for these units were mapped due to a lack of information. The formations that comprise these units, such as the nonmarine sandstone of the Eocene age Roslyn and Swauk Formations, which are separated by the Teanaway Basalt, and the interbedded sandstone-basalt Naches Formation are as much as 5,000 ft thick (Kinnison and Sceva, 1963; Campbell, 1989). The combined lateral extents of these units where they abut the CRBG are about 1,520 mi². Within this area, these units either crop out or underlie unconsolidated and (or) consolidated deposits. The Mesozoic and Tertiary units also underlie the CRBG and the Quaternary bedrock unit overlies either the CRBG or the Mesozoic and Tertiary units.

The extents of the surficial hydrogeologic units are shown in figure 8. Note that the basin-fill units in the structural basins are aggregated for display purposes (detailed extents of these units and their mapped thicknesses are shown in Jones and others (2006)). Deposits that are similar to the basin-fill deposits outside of the six structural basins are included on figure 8 to show their limited extents. These deposits are considered part of the underlying hydrogeologic unit and are not important components of the regional groundwater-flow system. Information on the area and thickness of the units and their components (table 1) and the correlation chart (table 2)

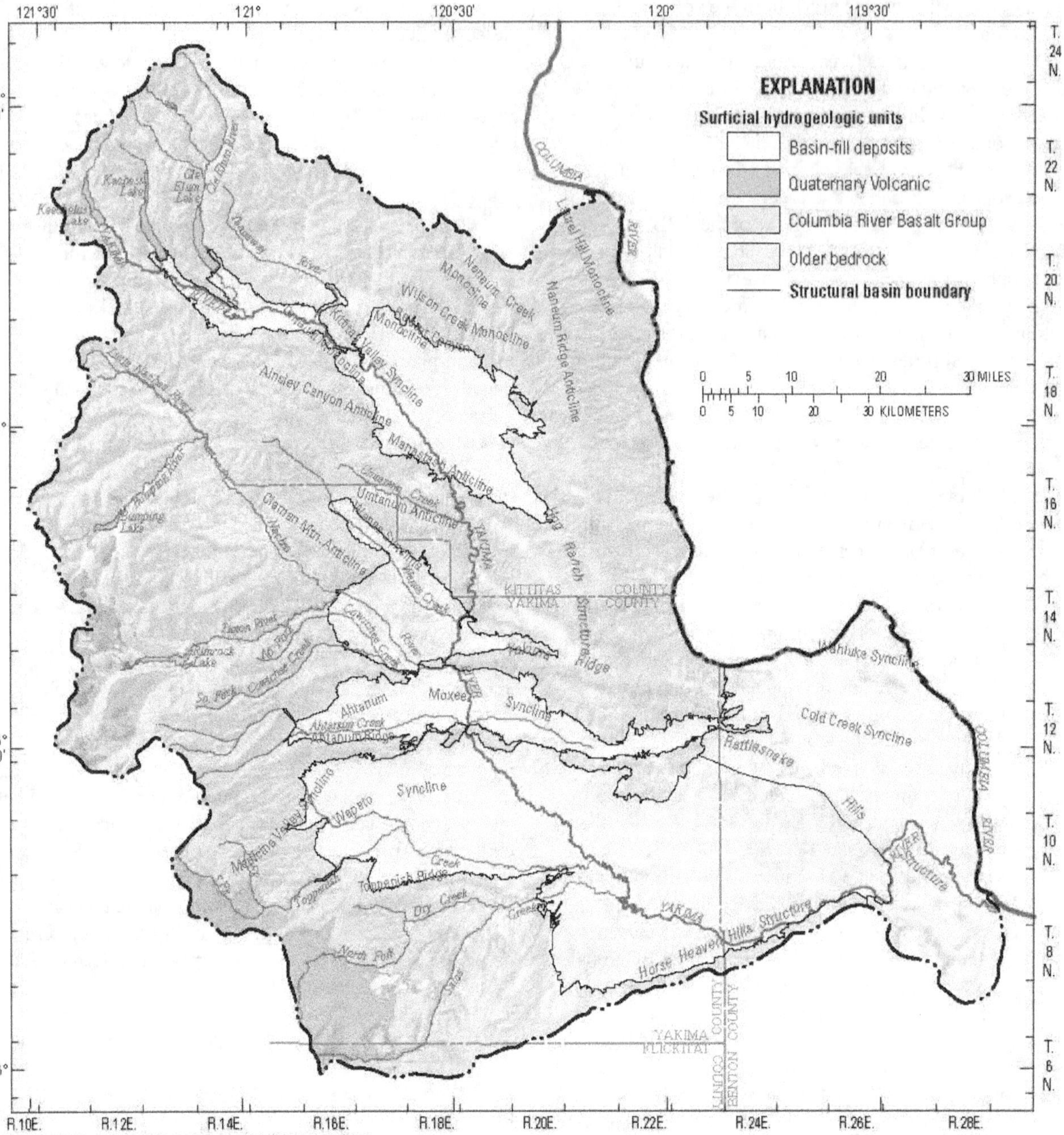

Base from U.S. Geological Survey and (or) Washington Division
of Geology and Earth Resources digital data, 1983, 1:100,000
Universal Transverse Mercator projection, Zone 10
Horizontal Datum: North American Datum of 1927 (NAD 27)
Vertical Datum: North American Vertical Datum of 1988 (NAVD 88)

Figure 8.   Extent of surficial hydrogeologic units, Yakima River basin aquifer system, Washington.

indicate the complex make-up of the Yakima River basin aquifer system.

Numerical Simulation of Groundwater Flow

Development of a calibrated groundwater-flow model allows for an analysis of the movement of water through the mapped hydrogeologic units that constitute the Yakima River basin aquifer system, and the potential simulated effects of stresses (and changes in stresses) on the flow system and surface-water resources. The U.S. Geological Survey modular three-dimensional finite-difference groundwater-flow model, MODFLOW-2000 (Harbaugh and others, 2000) was used to simulate groundwater flow in the basin-fill deposits, basalt units, and bedrock units of the aquifer system, and the interaction of the groundwater-flow system with surface water. MODFLOW uses datasets describing the mapped hydrogeologic units, boundary conditions, pumpage, recharge, initial conditions, and hydraulic properties, and calculates hydraulic heads at discrete points (nodes in a model cell) and flows within the model domain. The model also was constructed to simulate streamflow using the streamflow routing capability of the model for much of the Yakima River and 17 river miles of the Naches River. The routing also accounts for 109 diversions and 4 return flows. The constructed model using these datasets is a transient model calibrated to the period October 1959 to September 2001 (water years 1960–2001; herein, a stated year is a water year [period from October 1 through September 30]), and is termed the existing-conditions model (ECM). Water year 1960 was selected as a suitable starting point because 1960 was an average year for precipitation and streamflow, and reservoir operation and management was not representative of the rest of the simulation period, groundwater development was minimal, and there is a paucity of data before 1960. The ECM is suited for providing information on the effects of regional stresses on the groundwater-flow system during this time period.

Model Hydrogeologic Units

The three-dimensional digital hydrogeologic framework developed for the ECM primarily is based on data used by and digital maps developed by Jones and others (2006) and Jones and Vaccaro (2008): Digital Elevation Model (DEM), geologic and hydrogeologic maps, cross sections, and lithologic well logs. The electronic data were assembled into a three-dimensional, spatially distributed hydrogeologic

representation using GIS for incorporation into the ECM. The three-dimensional representation was compared to the published information and adjusted where appropriate. An effort was made to honor the published maps (Jones and others, 2006; Jones and Vaccaro, 2008) so the model construction was as representative as possible. The regional scale of the groundwater model created some discrepancies, but the method described in this section created a reproducible, hydrogeologic representation in the ECM.

The representation includes model HGUs that are the basis for assigning hydraulic properties to the cells of the model grid using the Hydrogeologic-Unit Flow (HUF) Package (Anderman and Hill, 2000). The HUF package facilitates the discretization of the complicated geometry of the HGUs because the model HGUs are defined and assigned to grid cells in the HUF package. The HUF package allows for additional model HGUs or model layers to be added or subtracted without a complete re-formulation of the model input. Some grid cells are filled by a single model HGU; other grid cells contain multiple model HGUs. The HUF package calculates the hydraulic properties for each cell based on the model HGU properties occurring in the grid cell, thus, allowing the vertical geometry of the model HGUs to be independent of the finite-difference-model grid layers.

The 28 mapped HGUs were subdivided to define 48 model HGUs for the ECM (table 3). These model HGUs included 20 basin-fill units, 5 repeating sequences of Saddle Mountains unit interflow zones (generally high horizontal conductivity) and interiors (generally low horizontal hydraulic conductivity) (10 total), 3 sequences of Wanapum unit interflow zones and interiors (6 total), 4 sequences of Grande Ronde unit interflow zones and interiors (8 total), 2 interbeds, and 2 bedrock units (upper 10 ft of bedrock and remaining bedrock thickness). A representative cross section showing model HGUs located in the Toppenish basin is shown in figure 9.

The physical characteristics of the basalt flows govern the movement of groundwater in basalts, and thus its availability. Upper zones of the flows were exposed to weathering processes and were broken by subsequent flows, resulting in the formation of conductive "flow tops." These flow tops, when combined with the base of the overlying basalt flow, form interflow zones that readily transmit water (Lindolm and Vaccaro, 1988). The interflow zones commonly make up 5–10 percent of the thickness of a flow. The interflow zones are separated by the less transmissive entablature and colonnade in which the fractures typically are vertically oriented (Tomkeieff, 1940; Waters, 1960; MacDonald, 1967; Swanson and Wright, 1978; Sublette, 1986; Hansen and others, 1994). The fractures are a result of contraction during cooling of basalt flows (MacDonald, 1967; Long and Wood,

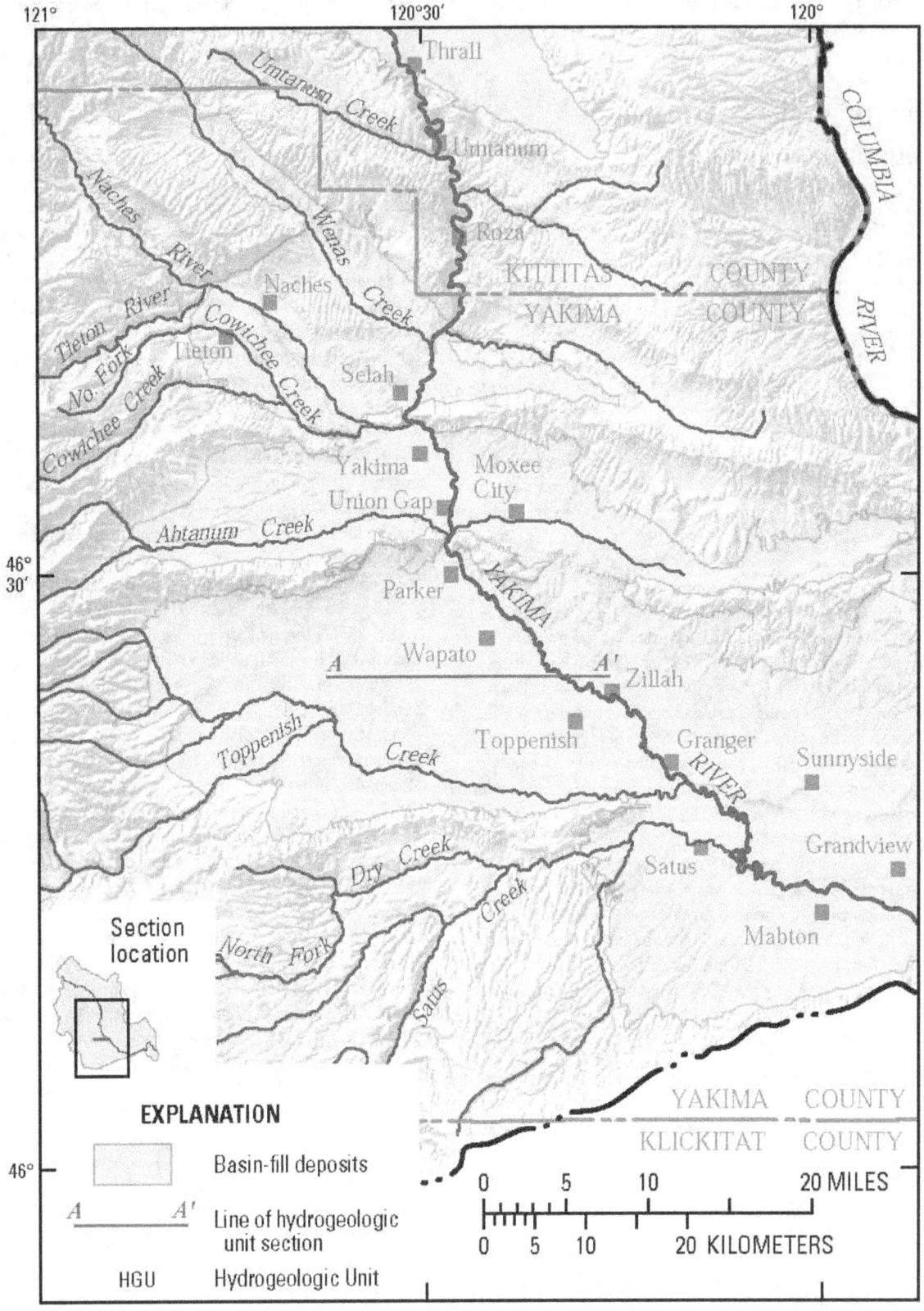

Figure 9.   Relation of model hydrogeologic units to model layers, Yakima River basin aquifer system, Washington.

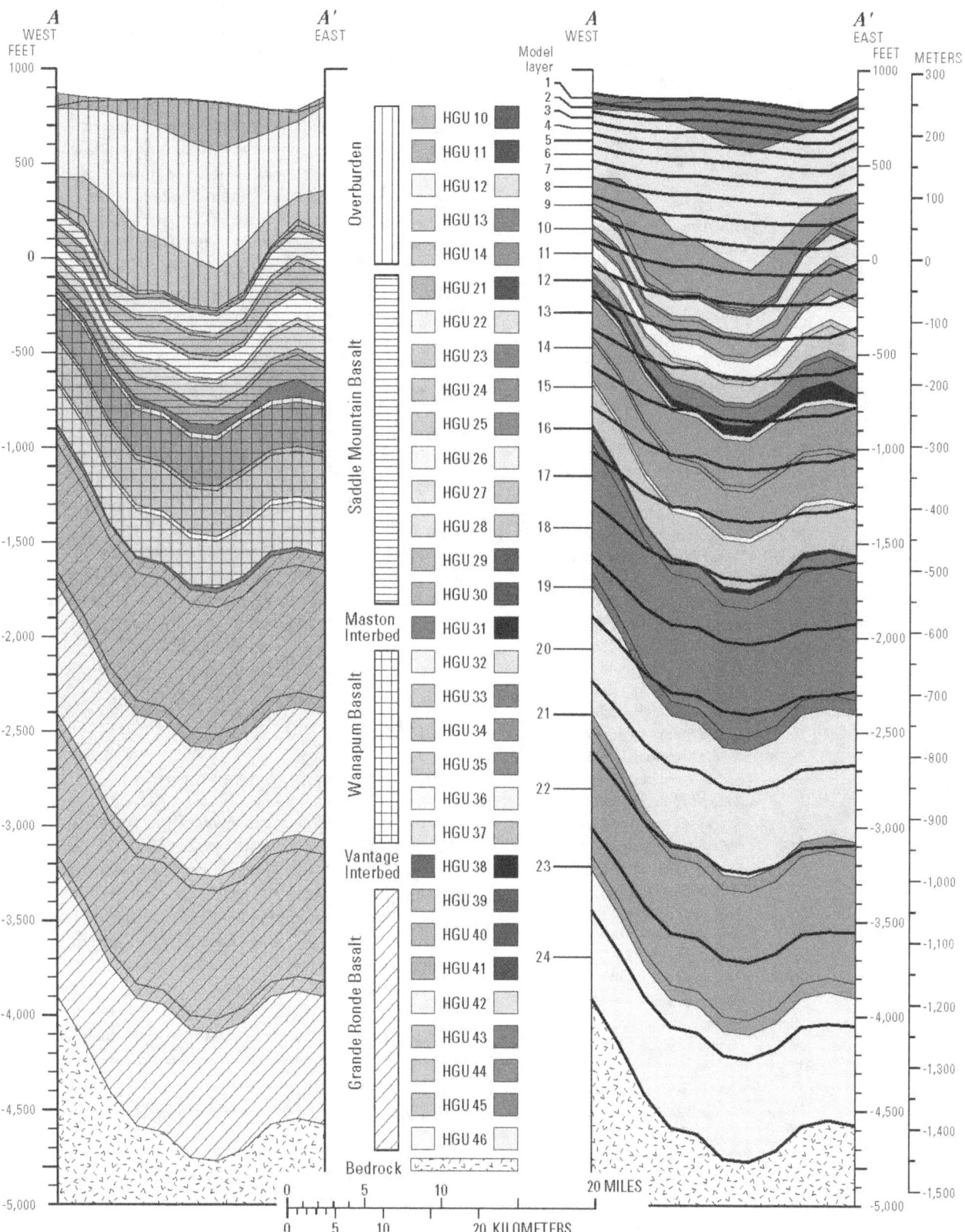

Figure 9.—Continued

Table 3. Information for the model hydrogeologic units, Yakima River basin aquifer system, Washington.

Structural basin	Model hydrogeologic unit	Mapped hydrogeologic unit
Roslyn	1	Upper aquifer
Roslyn	2	Clay
Roslyn	3	Basal gravel
Kittitias	4	Alluvium
Kittitias	5	Unconsolidated
Kittitias	6	Consolidated
Selah, Yakima	7	Alluvium
Selah, Yakima	8	Unconsolidated
Selah, Yakima	9	Consolidated
Toppenish	10	Fine-grained unconsolidated
Toppenish	11	Coarse-grained consolidated
Toppenish	12	Consolidated - top of Ellensburg
Toppenish	13	Consolidated - top of Rattlesnake Ridge "Blue clay"
Toppenish	14	Consolidated - base of Rattlesnake Ridge "Gravels"
Benton		Touchet - fine grained
Benton	16	Pasco gravels - coarse grained
Benton	17	Upper Ringold - fine grained
Benton	18	Middle Ringold - coarse grained
Benton	19	Lower Ringold - fine grained
Benton	20	Basal Ringold - coarse basal gravels
Regional	21	Saddle Mountain - interflow
Regional	22	Saddle Mountain - interior
Regional	23	Saddle Mountain - interflow
Regional	24	Saddle Mountain - interior
Regional	25	Saddle Mountain - interflow
Regional	26	Saddle Mountain - interior
Regional	27	Saddle Mountain - interflow
Regional	28	Saddle Mountain - interior
Regional	29	Saddle Mountain - interflow
Regional	30	Saddle Mountain - interior
Regional	31	Mabton interbed
Regional	32	Wanapum - interflow
Regional	33	Wanapum - interior
Regional	34	Wanapum - interflow
Regional	35	Wanapum - interior
Regional	36	Wanapum - interflow
Regional	37	Wanapum - interior
Regional	38	Vantage interbed
Regional	39	Grande Ronde - interflow
Regional	40	Grande Ronde - interior
Regional	41	Grande Ronde - interflow
Regional	42	Grande Ronde - interior
Regional	43	Grande Ronde - interflow
Regional	44	Grande Ronde - interior
Regional	45	Grande Ronde - interflow
Regional	46	Grande Ronde - interior
Regional	47	Bedrock - upper 10 feet
Regional	48	Bedrock

1986) and of later folding and faulting. The greatest density of fractures generally occurs in the entablature (Wood and Fernandez, 1988; Reidel and others, 2002).

The repeating sequences for the basalt units were defined in order to regionally capture the vertically cyclical nature of interflow zones and interior flow zones. The interflow zones (basalt-flow top combined with overlying flow bottom) support much of the lateral flow that supplies water to wells, whereas interior flow zones (referred to here as flow *interiors*) are much tighter, and the vertical flow component is likely of the same order of magnitude as the lateral component. The number of sequences roughly accounts for each unit's named members, which are not mapped. Therefore, these repeating sequences account for the vertical variations, but are arbitrarily assigned to a thickness distribution based on the assumption that interflow zones account for about 10 percent of a basalt flow (Swanson and others, 1979c; Hansen and others, 1994). For example, the thickness of the Wanapum unit was divided into three subunits of equal thickness, and each subunit was then subdivided into two units with the interflow zone accounting for 10 percent of that thickness. Compared to previous model investigations on the Columbia Plateau where each unit has been modeled as an effective unit with no variations with depth (Davies-Smith and others, 1989; Lum and others, 1990; Hansen and others, 1994; Packard and others, 1996; Bauer and Hansen, 2000), this subdivision accounts for known lateral and vertical variations in hydraulic characteristics and thus, is an important control on the vertical movement of water in the basalt units.

Discretization

Spatial Discretization and Layering

In MODFLOW, the groundwater-flow system is subdivided laterally and vertically into rectilinear blocks called cells. The averaged hydraulic properties of the material in each cell are assumed to be homogeneous. The ECM consists of a horizontal grid of 600 columns and 600 rows that are a uniform 1,000 ft per side (fig. 10). The uniform grid spacing was selected to reflect the regional scale of this study. The very small grid size relative to such a large system reflects the need for capturing the variations in the system over a small spatial scale, especially to better represent streams and agricultural drains in the basin. Note that the small size of the grid cells does not imply precision at that scale. The lateral extent of active cells, which is representative of all layers,

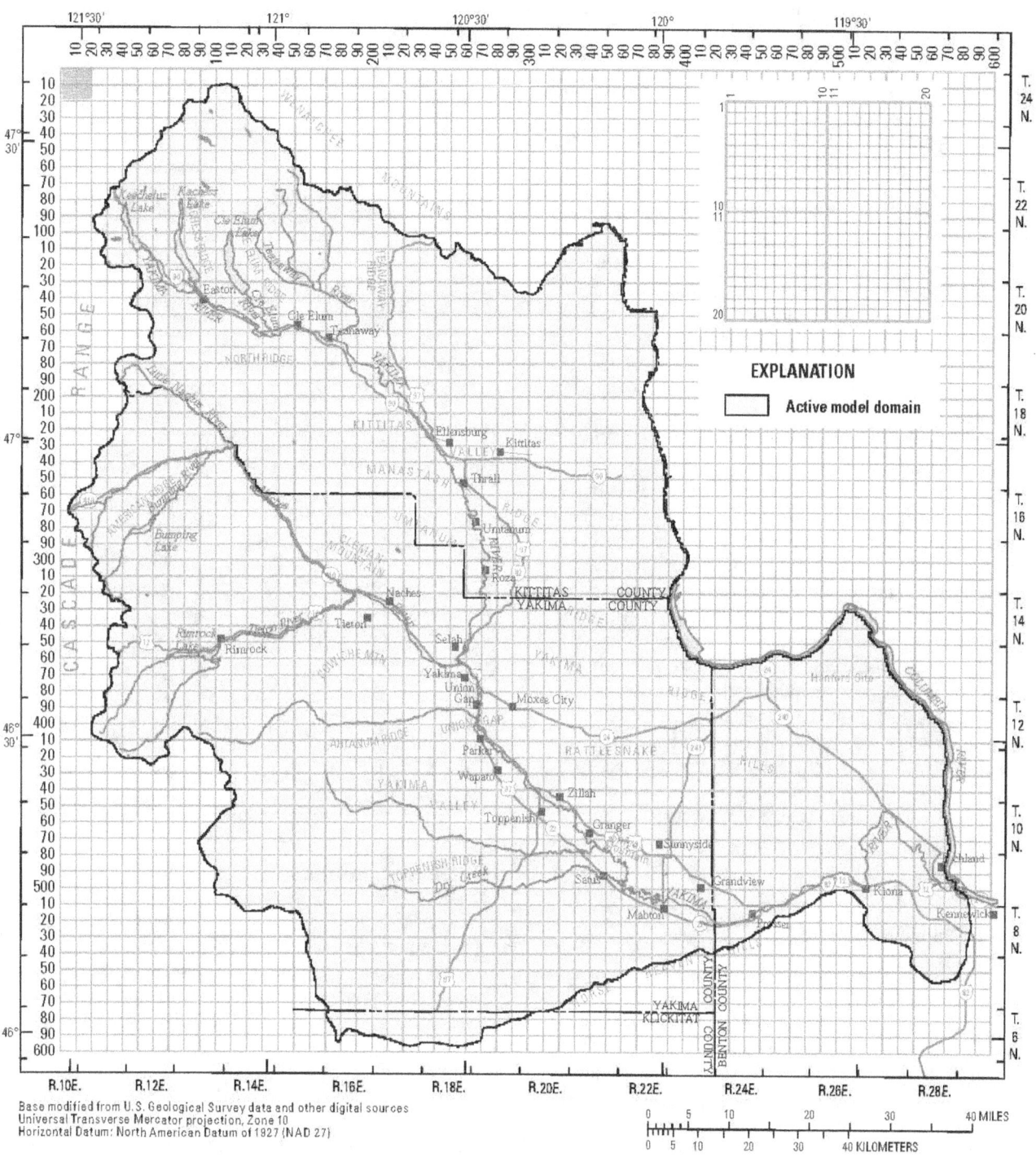

Figure 10.   Location and extent of the groundwater-flow model grid and the model hydrogeologic units, Yakima River basin aquifer system, Washington. The insert depicts the detailed horizontal discretization for the first 20 rows and columns of the grid.

and is the surficial extent of the model HGUs is shown in figure 10.

Vertically, the model domain was subdivided into 24 model layers, scaled vertically so cells become increasingly thick with depth (fig. 9); the combination of lateral and vertical discretization resulted in 8,640,000 cells within the model grid, of which 4,582,080 cells are active. The active cells include an area of 6,900 mi^2 and constitute 5,575 mi^3 of aquifer-system material. Total model domain thickness ranged from 1,780 to 8,800 ft. The percent thickness increase per layer increased parabolically with depth. Each cell in the top layer was specified as 1 percent of the total thickness at that row and column. Each successively lower layer was specified as a slightly larger percent of total thickness, with the lowest and thickest layer, 24, being 9.8 percent of the total thickness. Mean layer thickness for layer 1 is 43 ft with a maximum thickness of 88 ft, and mean layer thickness for layer 24 is 419 ft with a maximum thickness of 866 ft (table 4). The parabolic-layering approach was used in order to provide relatively finer vertical discretization for upper layers of the model, especially those in close proximity or connection to rivers or streams. The model extended to depths in excess of 8,000 ft, to assess potential impacts of pumping from deeper aquifers, but did not require as fine a discretized grid for simulation purposes as did near surface layers. Constructing the ECM with 24 model layers and a maximum thickness of 8,800 ft was considered necessary because questions concerning the management of the water resources occur at various scales throughout the model domain.

All model layers were simulated as confined because simulating any layers as convertible greatly increased numerical instability and computational time. Simulating all layers as confined introduces some inaccuracies and undoubtedly has some effect on the computed stream-aquifer interactions.

Temporal Discretization

The simulation period extends from October 1, 1959, to September 30, 2001, for a total of 42 water years (1960–2001) and 504 monthly stress periods, within which specified stress data are constant. The number of time steps in a stress period is equal to the number of days in the month. The monthly stress period and daily time step created efficient model-run times while still capturing flow-system characteristics at a sufficiently small temporal scale to meet study objectivesA large range in climatic conditions is explicitly included in this time period (water years 1960–2001) and thus, this time period includes changes in unregulated runoff, groundwater recharge, pumpage and reservoir operations. The long simulation period allows for a temporal assessment that accounts for these large changes in the flow system.

Table 4. Descriptive statistics for the thickness of the groundwater-model layers, Yakima River basin aquifer system, Washington.

[**Abbreviation:** ft, foot]

Model layer	Percent of total model thickness	Minimum layer thickness (ft)	Mean layer thickness (ft)	Maximum layer thickness (ft)
1	1.00	18	43	88
2	1.06	19	45	93
3	1.15	20	49	101
4	1.26	22	54	111
5	1.41	25	60	124
6	1.59	28	68	140
7	1.80	32	77	159
8	2.03	36	87	179
9	2.30	41	98	203
10	2.59	46	110	228
11	2.92	52	125	257
12	3.27	58	139	288
13	3.66	65	156	323
14	4.07	72	174	359
15	4.51	80	192	398
16	4.98	89	212	439
17	5.48	97	234	483
18	6.02	107	257	531
19	6.58	117	281	580
20	7.17	127	306	632
21	7.79	139	332	687
22	8.43	150	359	743
23	9.11	162	388	803
24	9.82	175	419	866

Boundary Conditions

Boundary conditions define the locations and manner in which water enters and exits the active model domain. The conceptual model for the aquifer system is that water: (1) enters the system as recharge from precipitation (rainfall and snowmelt) and recharge from the delivery and application of surface-water irrigation, and (2) exits the system as streamflow, evapotranspiration, and groundwater pumpage. The specified boundaries of the model coincide as much as possible with natural hydrologic boundaries. Four types of model boundaries were used in the ECM: no-flow boundaries (groundwater divides), head-dependent flux boundaries (drains, streams, reservoirs), specified head boundaries, and specified-flux boundaries (recharge, pumpage, and streamflow inflow).

No-Flow Boundaries

Major topographic divides (highs) coincide with the lateral model boundaries with the exception of the eastern boundary, which coincides with the Columbia River. The model boundaries that coincide with the natural landscape divides are simulated as no-flow boundaries as they are assumed to be groundwater divides. The topographic divides are either exposed bedrock or basalt. These divides are the crest of the Cascade Range to the west, the drainage-basin divide of the Yakima and Wenatchee Rivers to the north, and Horse Heaven Hills to the south (fig. 11).

Head-Dependent Flux Boundaries

Drain Conductances and Stages

The MODFLOW Drain (DRN) package was used to simulate the headwater streams of the Yakima River basin (fig. 11) and other selected streams (ephemeral and those with a very small mean annual discharge) throughout the model domain. This approach added numerical stability to the model in areas far from regions of greatest interest and accounted for the generally gaining, humid-headwater upland stream reaches. The simulated quantity of water exiting the system at a drain cell (MODFLOW only allows simulated groundwater flow into a drain cell) is equal to the product of a user-specified drain conductance and the difference between the simulated hydraulic head in the drain cell and the specified altitude of the drain (stream stage). Drain altitude was set at land surface (sampled from the USGS 10-m DEM) minus 10 ft. The drain hydraulic conductance is a function of the surrounding hydrogeologic material and the drain geometry. Information necessary to calculate a drain conductance, such as the distribution and hydraulic conductivity of material near the drain, were unavailable. Commonly, drain conductance is a lumped parameter that is adjusted during calibration to match measured flows. For this study, drain conductance initially was set at a uniform 500,000 ft²/d. In comparison, the calibrated Columbia Plateau regional aquifer system model (CPMod; Hansen and others, 1994) used a drain conductance ranging from 43,000 to 950,000 ft²/d, with a mean value of 268,000 ft²/d. The ECM includes a total of 12,694 drain cells assigned to model layer 1.

Stream Conductances and Stages

The exchange of groundwater and surface water is an important hydrologic process in the groundwater-flow system and, to the extent possible for a regional model, the ECM was constructed to capture this process. Excluding drain cells described above, the Yakima River and its tributaries were simulated using the MODFLOW Streamflow-Routing (SFR2) package (Niswonger and Prudic, 2005) to route streamflow

and calculate river-aquifer exchanges. The ECM has 250 SFR2 segments and 8,533 reaches (cells); the locations of the SFR2 cells are shown in figures 11 and 12.

The exchange of water between the two systems is controlled by the differences between groundwater levels and stream stage. Stream stages for most tributaries to the Yakima and Naches Rivers were determined using the USGS 10-m DEM at various locations along the streams and stages were linearly interpolated between these locations. Some inaccuracy was introduced in the simulation of groundwater flow to and from the streams by using average stream stages and simulating average groundwater altitudes within model cells. This uncertainty was not deemed a problem in gentle relief areas, but uncertainty was introduced in areas of steep terrain and deeply incised canyons (with seepage faces contributing to streamflow). Many agricultural drain and drain systems were treated the same as stream tributaries. These drains include Wilson, Cherry, Wide Hollow, Sulphur, Spring, and Snipes Creeks, and Moxee, Marion, and Granger Drains (fig. 12); these agricultural drains also function as wasteways. The remaining smaller streams were assigned a depth of 2 ft and width of 20 ft. The altitude of the top of the streambed was calculated as the DEM elevation minus the stream depth, or in the case of the smaller streams land surface minus 2 ft.

For the Yakima and Naches Rivers, stream depth was computed using Manning's equation assuming an eight-point cross section. Average depth and width for the cross sections were based on mean annual streamflow from the USGS National Hydrography Dataset (http://nhd.usgs.gov/index.html) and regression equations determined by Magirl and Olsen (2009).

Streambed thickness was set at 1 ft for all stream reaches. For the parts of tributary streams in steep upland areas, a depth of 20 ft was used. This assumed depth accounts for the water table intersecting the valley walls above the stream (Winters and others, 1998).

The simulated quantity of water moving between the groundwater and surface-water systems is equal to the product of streambed conductance and the simulated head difference between the stream and underlying model HGUs. Initial values of streambed conductance were based on stream length (determined using GIS) and width (Magirl and Olsen, 2009), estimated streambed hydraulic conductivity, and streambed thickness. Initial estimates of streambed hydraulic conductivity were based on Hansen and others (1994) and adjusted during model calibration. The model internally multiplies the hydraulic conductivity value (ft/d) by the stream reach length (ft) and width (ft), divided by the streambed thickness (ft), resulting in the streambed conductance (ft²/d). For routing streamflow, a constant value of 0.04 was used for Manning's coefficient in streams where depth and stage are calculated by the model.

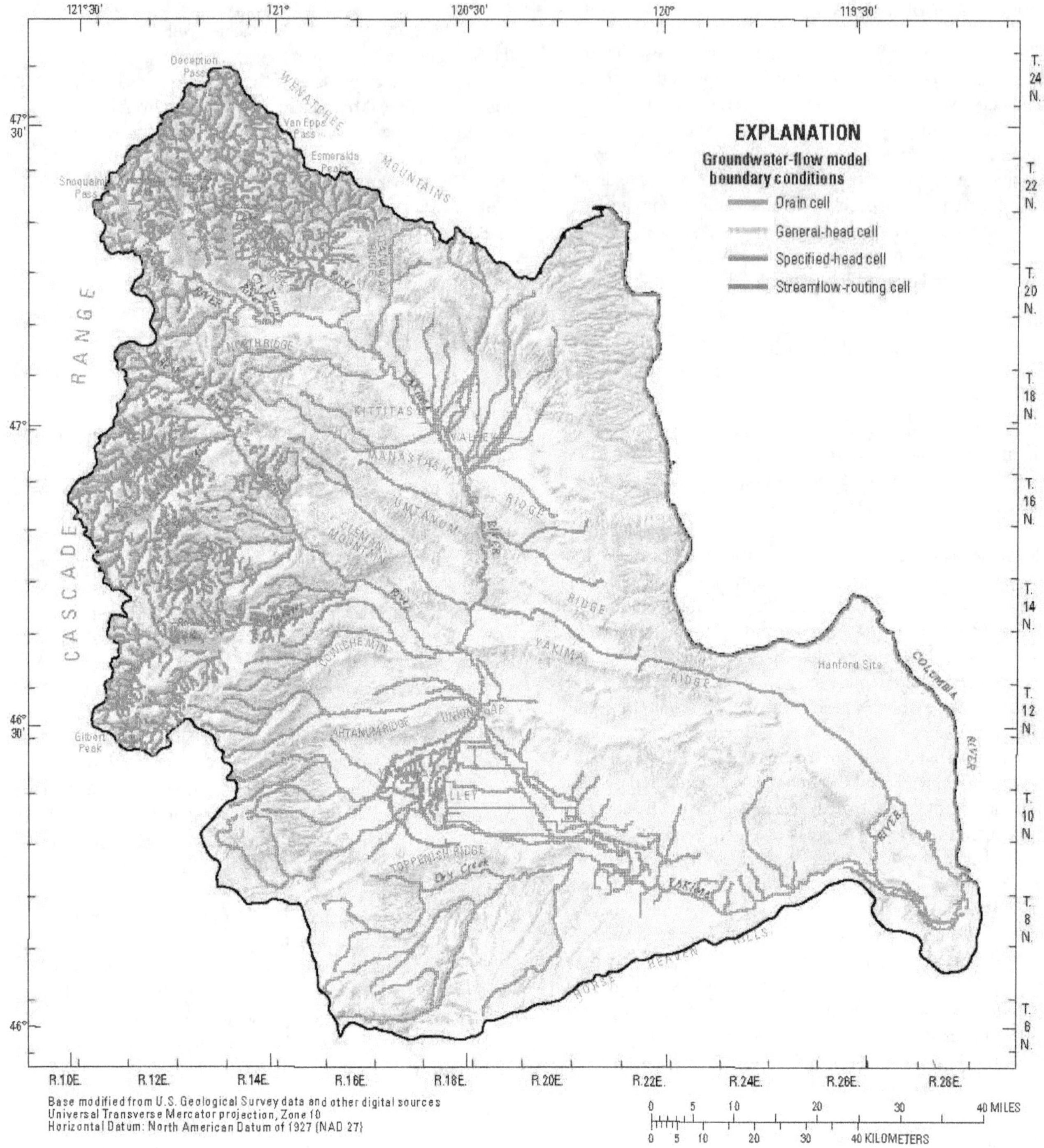

Figure 11.   Location of the groundwater-flow model boundary conditions for drain, general-head boundary, specified-head boundary, and streamflow-routing cells, Yakima River basin aquifer system, Washington.

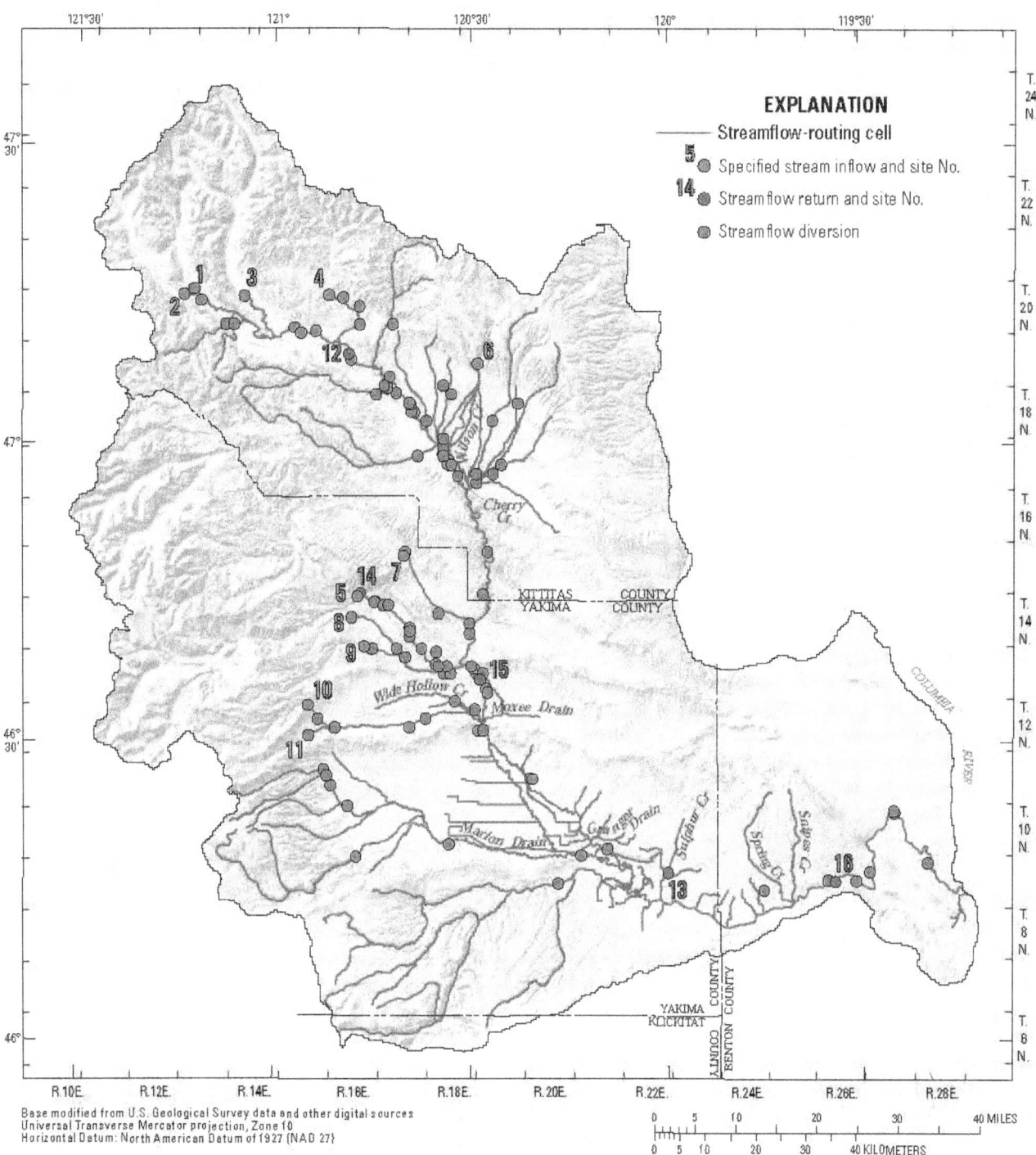

Figure 12. Location of streamflow-routing cells and points of diversions, returns, and specified stream inflows, Yakima River basin aquifer system, Washington.

General-Head Boundary Conductances and Stages

The MODFLOW general-head boundary (GHB) package was used to simulate subsurface exchange between five reservoirs (Keechelus, Kachess, Cle Elum, Bumping, and Rimrock Lakes—see figs. 1 and 11) and the underlying model HGUs. The ECM includes 880 general-head boundary cells that are all assigned to model layer 1.

An external source (lake) provides flow into or out of a cell within the groundwater system in proportion to the difference between the simulated hydraulic head in the cell and the specified head (lake stage) of the external source. The specified lake stages were determined from USGS 1:24,000-scale topographic maps and held constant throughout the simulation period. The lake-bottom conductance is a function of the surrounding hydrogeologic material and the lake area. Information about lake conductance was unavailable, as it was for drain conductance. The lakebed conductivity was initially set at 1 ft/d and multiplied by the lake area in the model cell divided by a 1-ft lakebed thickness.

Specified-Head Stages

The MODFLOW specified-head (CHD) package was used to simulate the Columbia River (the eastern boundary of the model; fig. 11). Monthly streamflow for USGS gaging station, Columbia River below Priest Rapids Dam (12472800) during the simulation period ranged from 52,500 to 461,000 ft³/s. Specified head at each cell was based on the stage-discharge relation for the Hanford Reach of the Columbia River at mean annual flow simulated with a two-dimensional hydraulic model (T. Waddle, U.S. Geological Survey, written commun., 2008). Behind the few dam locations along this boundary, head was set at mean pool elevation. Mean differences between monthly pool elevations over the simulation period were less than 5 ft and most differences were less than 3 ft.

Specified-head boundaries were used at 748 surficial cells, and were assigned to all 24 vertical model layers at those locations. This resulted in a total of 17,952 CHD cells. Measured water levels near the Columbia River increased with depth. To represent this measured upward gradient, heads at each successively lower layer were assigned a value 1 ft higher than the head assigned to the overlying model layer, thereby forcing upward flow at the model boundary. The 1-ft value was used because it approximates the measured water levels from a 4,800-ft deep well near the Columbia River on the Hanford site that are on the order of 15–20 ft above river level. Some unknown quantity of groundwater may flow beneath the Columbia River, but the general conceptual understanding of the flow system is that the Columbia River is the predominant regional sink of the system. This conceptual model is consistent with previous work, including that of Davies-Smith and others (1989), Drost and others (1990), Packard and others (1996), and Hansen and others (1994). The main purpose of the boundary was to extend the model domain well beyond the Yakima River basin to diminish any Columbia River boundary effects in the ECM in the basin. Throughout most of the area, minimal pumping occurs near the boundary (less than 0.01 percent of the total pumpage), further diminishing possible boundary effects.

Specified-Flux Boundaries

Fluxes were specified in the model for the recharge to and groundwater withdrawals from the groundwater system and inflow to 11 upstream SFR2 segments. Because there are no major interbasin transfers of water, groundwater recharge, primarily from snow and rain, is the primary simulated specified-flux input, and groundwater withdrawals are the simulated specified-flux outputs. Evapotranspiration was accounted for in the estimation of groundwater recharge (Vaccaro and Olsen, 2007).

Groundwater Pumping

Groundwater pumping in the Yakima River basin was estimated as part of this study for eight categories of use for 1960–2001 (Vaccaro and Sumioka, 2006). The eight categories of pumping were public water supply (including wells for Group B, systems defined as supplies with less than 15 connections), self-supplied domestic (permit-exempt wells), irrigation, frost protection, livestock and dairy operations, industrial and commercial processes, fish and wildlife propagation, and groundwater claims. Pumpage estimates were based on methods that varied by the category and primarily represent pumpage associated with groundwater rights. Methods, pumpage estimates, reliability of the estimates, and a comparison with appropriated quantities are described by Vaccaro and Sumioka (2006) and Vaccaro and others (2009). Total annual pumpage values in the ECM generally are consistent with pumpage values listed as "cumulative" to indicate total annual pumpage at the end of 5-year intervals in tables from these previous reports. Pumpage values will not be exact matches due to rounding of pumpage quantities, later identification that a standby/reserve well was a primary irrigation well, elimination of some wells with small pumpage quantities, and obtaining additional information on the pumping rates for some wells late in the study. Pumpage estimates are described in more detail in section, "Model Applications." Total withdrawals and number of withdrawal points, by stress period are shown in figures 13A and 13B.

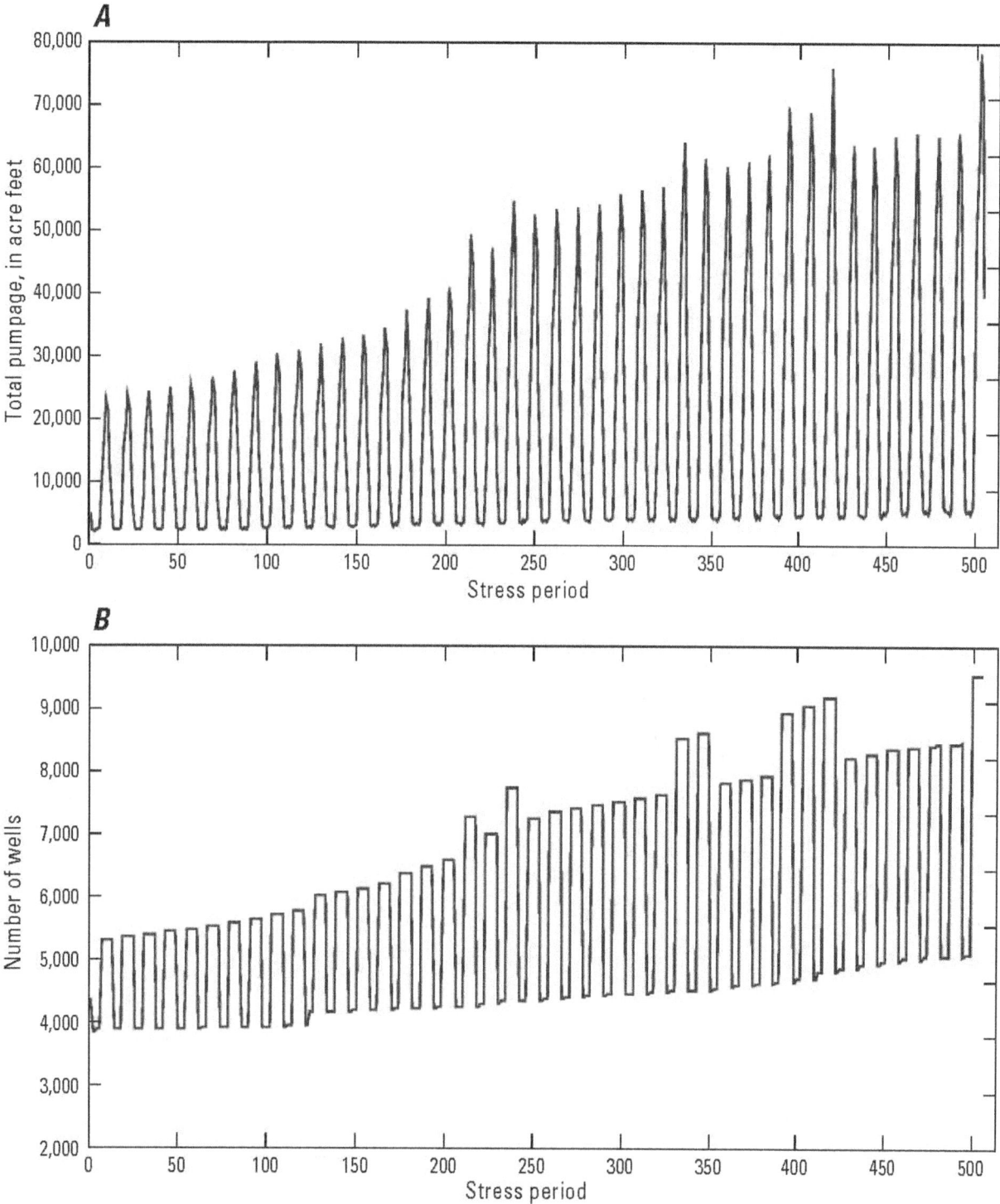

Figure 13. Total pumpage and number of wells, by stress period, and number of wells by model layer for July 2000 and 2001, and percentage of the total pumpage in July 2000 by model layer, Yakima River basin aquifer system, Washington.

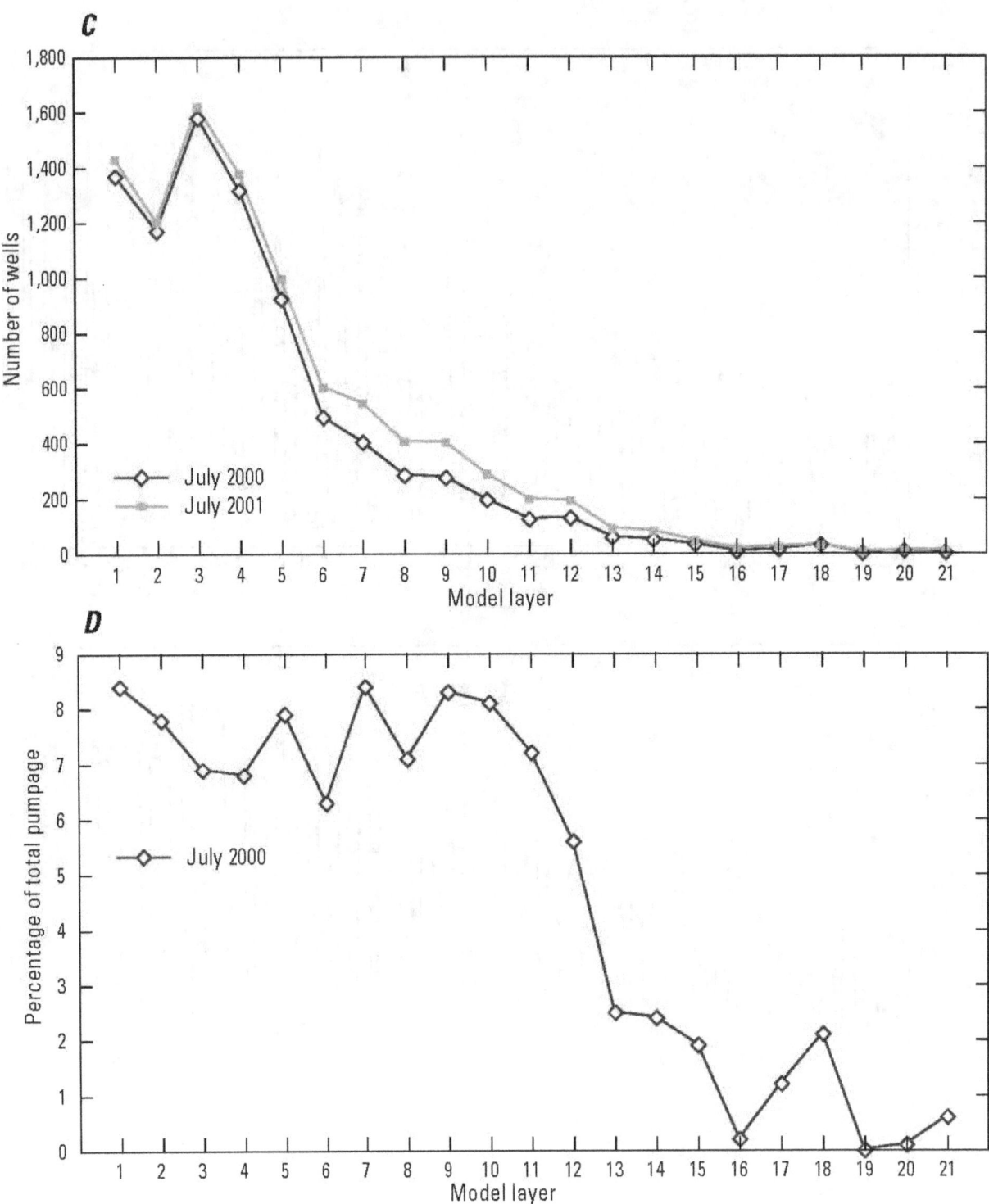

Figure 13.—Continued

By the end of calendar year 1960, total annual pumpage in the basin, excluding standby/reserve pumpage, was about 116,000 acre-ft (160 ft³/s). By 2000, total annual pumpage (without standby/reserve rights), was estimated to be about 317,000 acre-ft (440 ft³/s). Irrigation accounts for about 60 percent of the pumpage, followed by public water supply at about 12 percent. The smallest category of pumpage was for livestock/dairy use with pumpage in 2000 estimated to be about 6,700 acre-ft (10 ft³/s). However, total pumpage for this category is not well understood because under Washington law, it is permissible to pump any quantity necessary to meet stock-watering needs and no information is available on the withdrawal quantities. Thus, locally the livestock/dairy use may be underestimated by an unknown factor.

Total annual pumpage in 2000 was about 11 percent of the surface-water demand and about 12 percent of the Yakima River annual mean streamflow leaving the basin at Kiona. Maximum pumpage is in July and August, and during the non-prorating year of 2000, a year of normal recharge, the maximum pumpage was about 90 and 80 ft³/s, respectively; in contrast, July pumpage in the dry, prorating year of 2001 increased to about ·110 ft³/s.

The ECM simulates groundwater pumpage at a maximum of 9,558 modeled points of withdrawal during the 2001 irrigation season; the minimum number of points of withdrawals (3,877) occurred in November and December 1959. The assignment of the pumpage to the model layers is described in appendix A. The distribution of wells by model layer for July 2000 (average year) and 2001 (dry year) with additional irrigation groundwater withdrawals (fig. 13C), shows that most points of withdrawal are in the upper part of the system, with about 64 and 59 percent in layers 1–4 during July 2000 and 2001, respectively. For July 2000 and July 2001, 75 and 69 percent of the points of withdrawals, respectively, were from layers 1–5. Pumpage from model layers 1–7 for July 2000 and July 2001 comprises 81 and 86 percent, respectively, of all withdrawals. These points of withdrawals accounted for only about 53 and 51 percent of the total pumpage, indicating that the remaining 19 and 14 percent of the points of withdrawals in the deeper model layers accounted for nearly half the total pumpage in July 2000 and July 2001, respectively. These deeper wells typically are large-capacity irrigation and municipal wells. The percentage distribution of the total pumpage in July 2000 is shown by model layer in figure 13D.

Groundwater Recharge

Process-based models that compute distributed water budgets on a watershed scale have been used successfully to calculate groundwater recharge rates at varying spatial-temporal scales using readily available databases. Two models in the U.S. Geological Survey's Modular Modeling System (Leavesley and others, 1996) were used to estimate recharge: the Precipitation-Runoff Modeling System (PRMS) (Leavesley and others, 1983) and Deep Percolation Model (Bauer and Vaccaro, 1987; Vaccaro, 2007). The two models were used to simulate two different hydrologic regimes of the Yakima River basin (Vaccaro and Olsen, 2007a), and to estimate groundwater recharge for predevelopment and current land-use and land-cover conditions for those regimes (fig. 14).

PRMS is a process-based, deterministic, distributed-parameter model designed to analyze the effects of climate and land use on streamflow and basin hydrology. PRMS is designed for mountainous, snow-dominated environments. Four operational, calibrated PRMS models were available for estimating daily recharge (Mastin and Vaccaro, 2002) throughout about 53 percent of the extended study area, and they were used to estimate recharge for 1950–2003 (Vaccaro and Olsen, 2007a). The Deep Percolation Model originally was developed as a tool for estimating daily groundwater recharge over a broad array of landscapes and spatial-temporal scales and is best suited for areas dominated by agricultural irrigation. As part of the overall study, Vaccaro and Olsen (2007a) constructed 17 Deep Percolation Models for the areas in the basin not included in the PRMS models, about 35 percent of the extended study area, and estimated daily recharge for 1950–2003.

The mean annual groundwater recharge for the 6,200 mi² basin for predevelopment and existing conditions was estimated to be about 12 in. (5,525 ft³/s or 4 million acre-ft) and 16.4 in. (7,460 ft³/s or 5.4 million acre-ft), respectively (Vaccaro and Olsen, 2007a). The increase in recharge from predevelopment to existing conditions is due to the delivery and application of irrigation water to croplands. The active model domain is a larger area than the area used for the original determination of groundwater recharge (Vaccaro and Olsen, 2007a) and therefore a method was needed to extend the estimates of groundwater recharge to the areas outside of the original basin within the model domain (Vaccaro and others, 2009). Mean annual groundwater recharge for the model domain during the simulation period (1960–2001) was estimated to be about 13.6 in. (6,950 ft³/s or 5.0 million acre-ft).

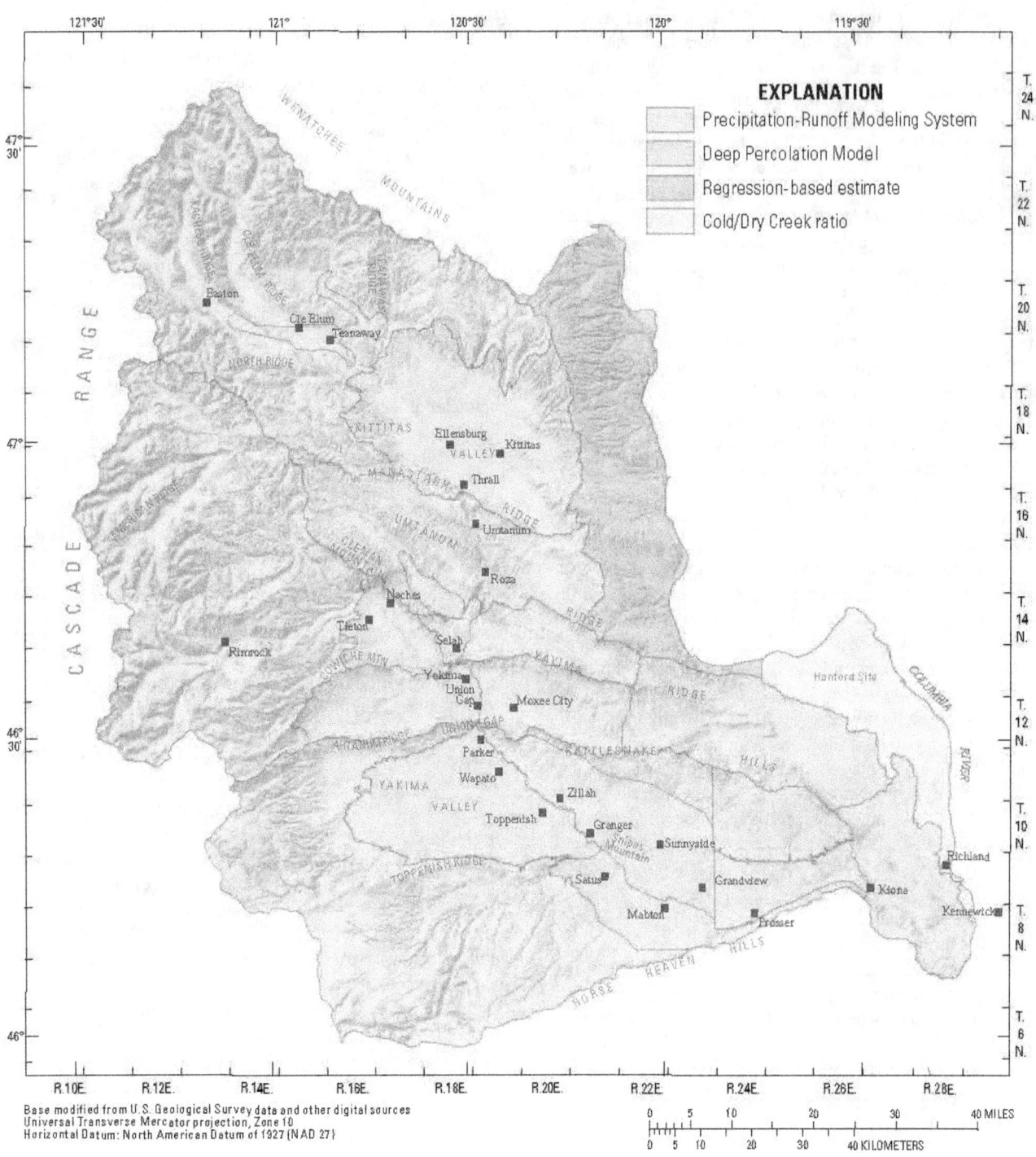

Figure 14.   Areas modeled to estimate groundwater recharge for current conditions, Yakima River basin aquifer system, Washington.

Mean monthly recharge rates estimated from the watershed models were used as initial input into the ECM and applied to the top model layer. This coupling of surface-water and groundwater models created a more physically deterministic representation of the flow system and allowed for an independent assessment of the recharge estimates. Initial model simulations indicated that several changes in the initial recharge estimates would result in improved simulations and a more realistic hydrologic framework. First, the estimated recharge for the irrigation districts was redistributed in time to increase recharge during the irrigation season and decrease it during the remaining months. The justification for this change was that though canal leakage was modeled in the previous watershed models as excess precipitation available for evapotranspiration, canal leakage actually occurs as a line source throughout the irrigation season. Estimated recharge in drier areas, especially those areas with a deep water table, was highly variable throughout the 42-year simulation period. These interannual and intermonth variations were expected to attenuate with depth and approach a constant value. Therefore, for the drier areas in the extended study area, outside of the river basin, the mean monthly recharge value (sum of monthly mean recharge for the same month for each year divided by the number of years) was used in place of the monthly mean recharge (sum of the daily mean recharge for one month divided by the number of days in that month). Furthermore, the minimum rate of recharge was assumed to be equivalent to 0.1 in/yr. This rate was applied to cells that had zero estimated recharge in those months. As a result of these changes, the estimated mean annual recharge used in the calibrated model was increased by 13 ft^3/s, from 6,950 to 6,963 ft^3/s (fig. 15).

Specified Stream Inflows

Monthly inflow to 11 upstream SFR2 segments (fig. 12) for the 504 stress periods was estimated from: (1) measured streamflow, (2) simulated streamflow from watershed models (Mastin and Vaccaro, 2002), and (3) a combination of measured data and regression equations derived from the measured data. Five of the 11 inflows account for more than 90 percent of the water entering the river system; these 5 inflows are based on measured data. The most upstream inflow was the Yakima River upstream of the confluence with the Kachess River (site 1), which was calculated using the measured monthly mean flow for the Yakima River at Easton minus measured monthly mean flow for the Kachess River near Easton plus measured monthly mean flow for the Kittitas Canal diversion. The next three major downstream inflows are the Kachess River near Easton (site 2) and the Cle Elum River near Roslyn (site 3), followed by the Teanaway River at Forks, near Cle Elum (site 4). The remaining major inflow is the measured monthly mean flow for the Naches River near Naches (site 5). The other prescribed inflows were for the upper Naneum (site 6) and Wenas Creeks (site 7), and

the North and South Forks of Cowiche (sites 8 and 9) and Ahtanum Creeks (sites 10 and 11). All other streamflows (for example, Satus Creek) or parts of streams (for example, the remaining part of Ahtanum Creek downstream of the forks) were calculated based on stream-aquifer interaction using the SFR2 package.

The ECM includes monthly aggregated streamflow diversions (109) and returns (5). The streamflow returns are Kittitas Reclamation District 1146 Drop (site 12), Sulphur Creek (site 13; surface-water component of wasteway), and the Wapatox (site 14), Roza (site 15), and Chandler-Prosser (site 16) power returns [fig. 12; diversion and return estimates are described by Vaccaro and others (2009)]. Sixty-five of the diversion quantities were based on Reclamation information, 65 on WADOE information, and 12 on Yakama Nation information; the WADOE diversions in the ECM generally represent aggregated points of withdrawal of several small diversions. The Reclamation information accounted for more than 90 percent of the total diversion quantities. Maximum diversions typically occur in July, and the July discharge for the 109 diversions ranged from as small as 0.7 ft^3/s to more than 2,000 ft^3/s. Total monthly mean diversion quantities (including power diversions) can exceed 8,500 ft^3/s. Excluding the large diversions, most of the monthly diversion quantities are estimates for some or all months for the simulation period. Any errors in these estimates would propagate as errors in simulated streamflow and stream-aquifer exchanges (and locally, hydraulic heads in the shallow groundwater system).

During initial simulations, some smaller creeks were simulated as going dry (having no flow). This result was most prevalent in the upper basin and was caused by inaccurate initial estimates of streamflow diversions that were larger than the simulated streamflow. Originally, mean monthly diversion estimates were based on the allowable acre-feet per year quantity that was distributed on the basis of the irrigation demand curve described by Vaccaro and others (2009). These quantities were larger than the simulated streamflow and greater than the watershed-model-estimated streamflow in the creeks. Testing showed that the maximum allowable instantaneous diversion (in cubic feet per second) in the data provided by WADOE for these smaller diversions provided more realistic diversion quantities that were more consistent with watershed-model-estimated streamflow in the creeks especially in August and September. The mean monthly (in contrast to varying monthly mean values for larger diversions) diversion quantities for the 65 aggregated diversions during the 42-year simulation period are least known in the northern areas for the smaller diversions. Completely capturing such diversion detail as described was beyond the scope of the study, especially considering the regional framework of the ECM and lack of 42 years of measured monthly data for all but the largest diversions.

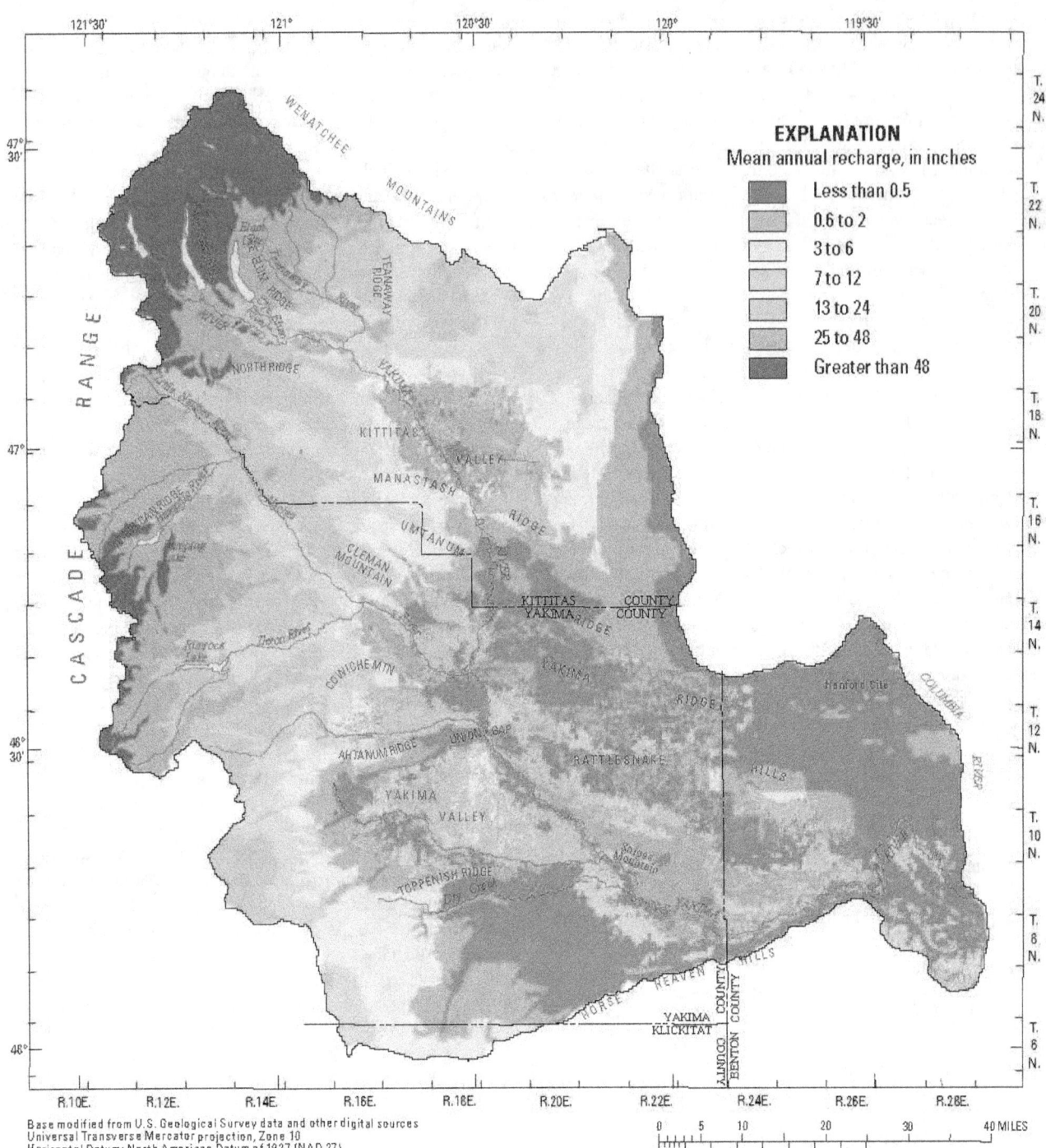

Figure 15. Mean annual recharge for current conditions, Yakima River basin aquifer system, Washington.

Initial Conditions

Initial conditions used in any transient model simulation are very important, as they serve as the boundary condition in time for the model simulation. They are discussed below based on Franke and others (1987) and Reilly and Harbaugh (2004); this discussion is nearly a verbatim, but shortened, version of their work. Initial conditions are the heads at the beginning of a transient simulation, and thus, serve as a boundary condition in time for the transient head response of the groundwater model solution. In transient solutions, the initial conditions are the heads from which the model calculates changes in the system due to the stresses that act on the system (in this study, pumpage and recharge). Thus, the response of the system is directly related to the initial conditions used in the simulation.

The changes in head that occur in the transient model due to any applied stress will be a combination of the effect of the change in stress on the system and any adjustments in heads as a result of errors in the initial head configuration (the initial conditions). Adjustments in heads resulting from errors in the initial head configuration do not reflect changes that would occur in the actual system, but occur because the heads specified as the initial condition are not a valid solution to the numerical model. Because errors in the initial head conditions cause changes in head over time during the simulation, it is best to begin all transient simulations with a head distribution that is a valid solution for the model.

For simulations that start from a period when the aquifer system was in steady-state equilibrium (meaning that the system remained unchanged for that period), the development of appropriate initial conditions is straightforward. In that case, a simulation of the steady-state period should be made and the results of this simulation should then be used as the initial conditions for the transient simulation. Sometimes, as is the case for the Yakima River basin aquifer system, it is not possible to start a simulation from a point in time where the aquifer was in steady-state equilibrium. This condition occurs because (1) the system is never at steady state due to varying climatic conditions and human influences and (2) it is not feasible to simulate the entire period of record from a time of a quasi-steady state because of resource constraints or lack of required information. Some rules of thumb for the evaluation of the appropriateness of the estimation of initial conditions in these non-ideal situations are to evaluate the time constant of the system under investigation and to test the sensitivity of the model to different initial conditions (Reilly and Harbaugh, 2004). For this study, both an extensive evaluation and testing were conducted.

A method was developed to derive initial ECM heads using transient simulations. Prior to water year 1960, groundwater-level changes due to pumping were minimal and only a few localized areas were affected. However, when large withdrawals commenced in 1960, this initial withdrawal stress propagated through the system as localized water levels tended to equilibrate. To account for the initial pumping in the transient model, the model was run repeatedly to simulate transient conditions using the 1960 monthly boundary conditions until head changes were minimal (1 ft) between initial 1960 and final 1960 heads. The resulting heads at the end of 1960 were assumed to be representative of any effects from historical pumping. However, some error associated with the initial heads is likely to exist and propagate as groundwater-level changes over time. The errors associated with initial heads and non-steady state conditions had little effect on the simulated streamflow for the Yakima and Naches Rivers, but introduced errors with some simulated streamflow for tributaries.

Reasonable initial conditions also resulted in dramatic improvement in numerical stability, a large decrease in computational time, and improvement in model fit. Development of initial conditions using this process was completed numerous times during model calibration, as changes to model parameter values changed the initial head distribution. This was especially true because of the transient effects of altering basalt hydraulic properties.

Initial Hydraulic Properties

The initial hydraulic properties of horizontal hydraulic conductivity (K_h), vertical hydraulic conductivity (K_v), anisotropy (K_h:K_v), and specific storage properties were assigned on the basis of values tabulated from both numerous previous studies and analysis of specific-capacity/aquifer-test data (Vaccaro and others, 2009), and on the basis of spatial distributions of these hydraulic properties in the calibrated CPMod (previously defined as the Columbia Plateau regional aquifer system model). It is important to understand the initial hydraulic properties because these are the values that are changed in model calibration, and some changes may indicate modifications to the overall conceptual model. Model HGUs are the basis for assigning K_h, K_h:K_v, and specific storage to the grid cells by the HUF package. For basin-fill (model HGUs 1–20) and the Vantage Interbed (model HGU 38) uniform hydraulic-property values were assigned, for the basalt (model HGUs 21–30, 32–37, and 39–49), arrays were used to account for the previously estimated spatial variations in the hydraulic properties in CPMod, and for the Mabton Interbed (model HGU 31) an array was assigned based on the fining of the mapped HGU to the east. Uniform values were used to simplify the model construction and calibration, and to regionalize hydraulic properties to the extent possible, which is appropriate for such a large regional model.

Hydraulic-property values were first assigned to model HGUs and the model was operated to test for solution criteria based on such aspects as numerical closure, model-budget error, and computational time. Model simulations were then completed with some trial-and-error adjustments to improve the spatial distribution of groundwater levels in layer 1. Parameter adjustments were not large. For example, K_h for a model HGU may have changed from 100 to 150 ft/d, and values were well within the ranges reported for mapped HGUs by Vaccaro and others (2009). These adjusted values are considered the initial hydraulic properties for calibrating the ECM.

Horizontal Hydraulic Conductivity

Horizontal isotropy was assumed for the basin-fill sediments, and each model HGU was initially assigned one value for K_h. The initial assigned values were estimated from K_h estimates for mapped HGUs by Vaccaro and others (2009). For the fine-grained HGUs, initial K_h values ranged from 0.5 ft/d for unit 2 (model HGU 2, lacustrine confining unit) in the Roslyn basin to 5 ft/d for the Touchet Beds (model HGU 10, deposited by Missoula Floods) in the Benton basin and the upper Ringold Formation (model HGU 19) in the eastern Benton basin (table 3). The coarse-grained units were assigned K_h values that ranged from 10 ft/d for the consolidated unit present in Kittitas, Selah, Yakima, and Toppenish basins (table 3) to 250 ft/d for the upper aquifer (unit 1, model HGU 1) in the Roslyn basin and the Pasco gravels in the eastern Benton basin (model HGU 16).

Initial hydraulic properties of the three basalt units follow the general trends of the properties from the calibrated CPMod that included all of the Saddle Mountains, Wanapum, and Grande Ronde units (Hansen and others, 1994). The K_h values for each unit for each of the CPMod model layers were overlaid on a map of the basin with topography, geologic structure, and groundwater levels. The gridded values from the CPMod were then smoothed by contouring values and accounting for geologic structure; smoothing was needed because of the relatively large size of the CPMod cells—about 1.5 by 2 mi. The contoured values were digitized and then interpolated to the ECM grid using GIS. The resulting values, which ranged from 0.5 to 30 ft/d, were used as the basis for the initial spatial distribution of K_h for both interflow zones and flow interiors (model HGUs 21–30, 32–37, and 39–48). The flow interior K_h values for the Saddle Mountains, Wanapum, and Grande Ronde units (model HGUs 21, 23, 25, 27, 29, 32, 34, 36, 39, 41, 43, and 45) were assigned an initial value that was based on a ratio of 10,000:1 (interflow zone K_h:flow interior K_h and the spatial distribution of K_h for the overlying interflow zones.

The initial estimates of K_h for the basalt interflow zones and flow interiors decreased with depth based on the method of Weiss (1982) and using the equation of Hansen and others

(1994). The assumption is that over time overburden pressure and secondary mineralization have reduced pore space with depth. The equation of Hansen and others (1994) reduces the K_h values with depth as a parabolic expression; the expression results in a 40-percent decrease of K_h at a depth of 3,000 ft. The method provided good estimates of K_h for the CPMod that included most of the Yakima River basin. The depth decay of K_h has been used by others in modeling groundwater flow in deep systems, for example, the Death Valley regional groundwater-flow system (Faunt and others, 2004). Faunt and others (2004) also found that a K_h decrease with depth was important for delineating the K_h of volcanic rock units, which may be hydrologically similar to the CRBG basalt units. In calibrating a regional groundwater-flow model for California's Central Valley, Faunt and others (2009) also found that depth decay of K_h was needed; maximum thickness in the model was on the order of 2,700 ft. The spatial distribution of the initial of K_h values for the upper interflow zones and flow interiors of the basalt units (fig. 16) shows a complex pattern and large differences between interflow zones and flow interiors. (The initial estimates K_h of the interflow zones and flow interiors were proportional, so the relative distributions of K_h values shown in figure 16 are the same but the values are many orders of magnitude different.)

Horizontal isotropy initially was assumed for the Mabton and Vantage Interbeds. The Mabton Interbed (model HGU 31) was assigned an initial K_h value of 0.25 ft/d, and excluding the basalt flow interiors, was the smallest of the initial K_h values. A K_h value of 10 ft/d was assigned to the Vantage Interbed (model HGU 38). The mapped HGU varies complexly over its extent, and locally the unit is used as a groundwater source. In some areas, the unit contains fine-grained materials. The Mabton Interbed K_h was then modified so that it varied spatially with the smallest values (0.25 ft/d) occurring in the eastern Benton basin where the unit is thickest and most fine-grained, and the largest values (1.0 ft/d) occurring to the west, where the unit is thinner and more coarse grained due to its proximity to source materials from the Cascade Range.

Horizontal isotropy also was assumed for the bedrock units, and each model HGU was initially assigned one value for K_h. A thin, upper part of the bedrock unit that was defined as model HGU 47 was assigned a value of 50 ft/d. This unit represents the regolith that contains mountainous coarse-grained soils, and locally, unconsolidated deposits derived from the Pleistocene glacial tills and conglomerates. Glacial loading likely also induced increased fracturing in this part of the system. The remaining part of the bedrock units (model HGU 48, more than 99 percent of the total bedrock volume) initially was assigned a K_h of 0.1 ft/d for the older bedrock and 10.0 ft/d for the younger Quaternary volcanics (fig. 8) based on a review of the literature survey for hydraulic properties of bedrock; the 0.1 ft/d value is near the highest of reported bedrock values (Vaccaro and others, 2009).

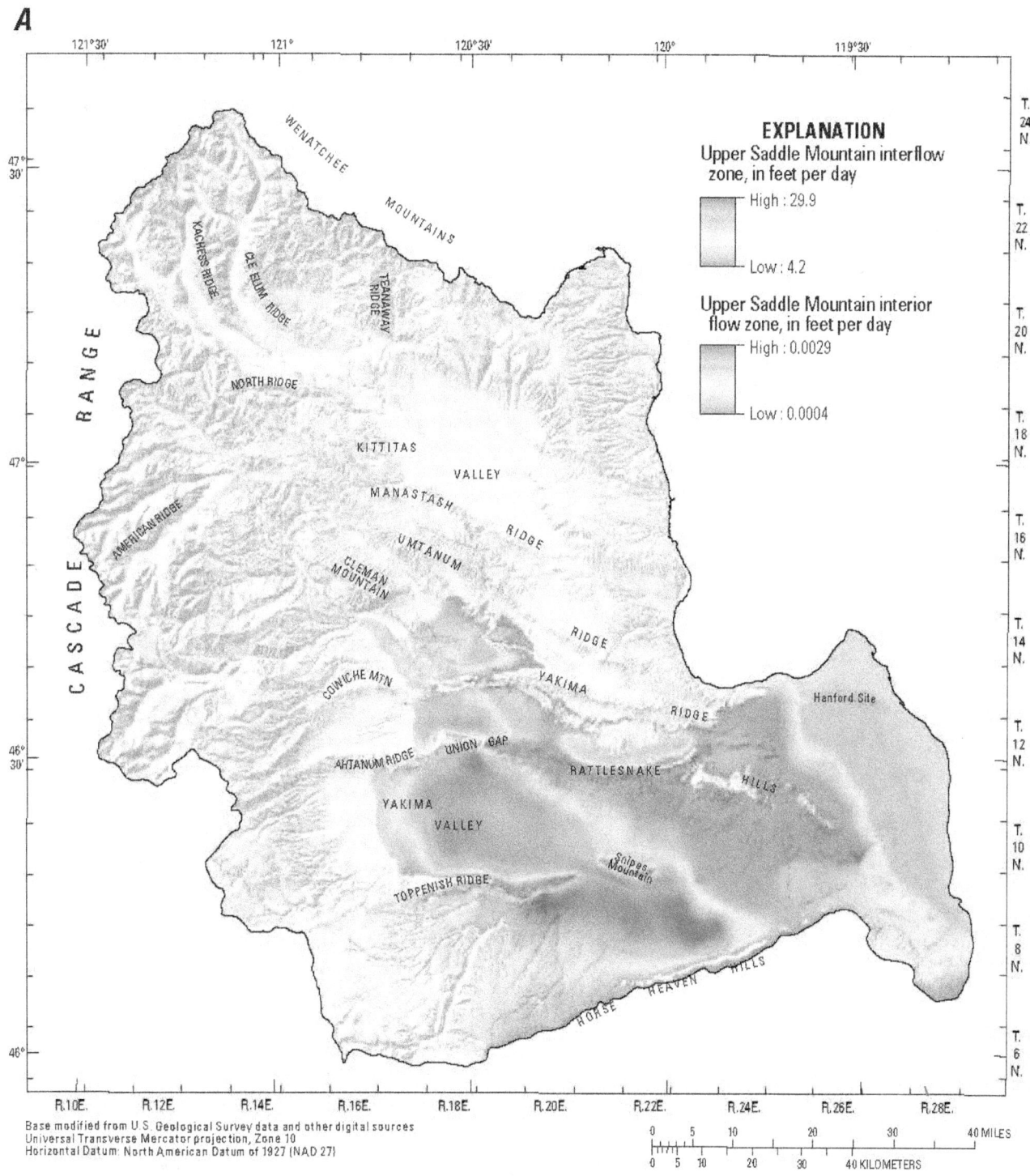

Figure 16. Spatial distribution of initial estimates of horizontal hydraulic conductivity for the interflow zones and flow interiors of the (*A*) upper Saddle Mountains, (*B*) upper Wanapum, and (*C*) upper Grande Ronde basalt units, Yakima River basin aquifer system, Washington.

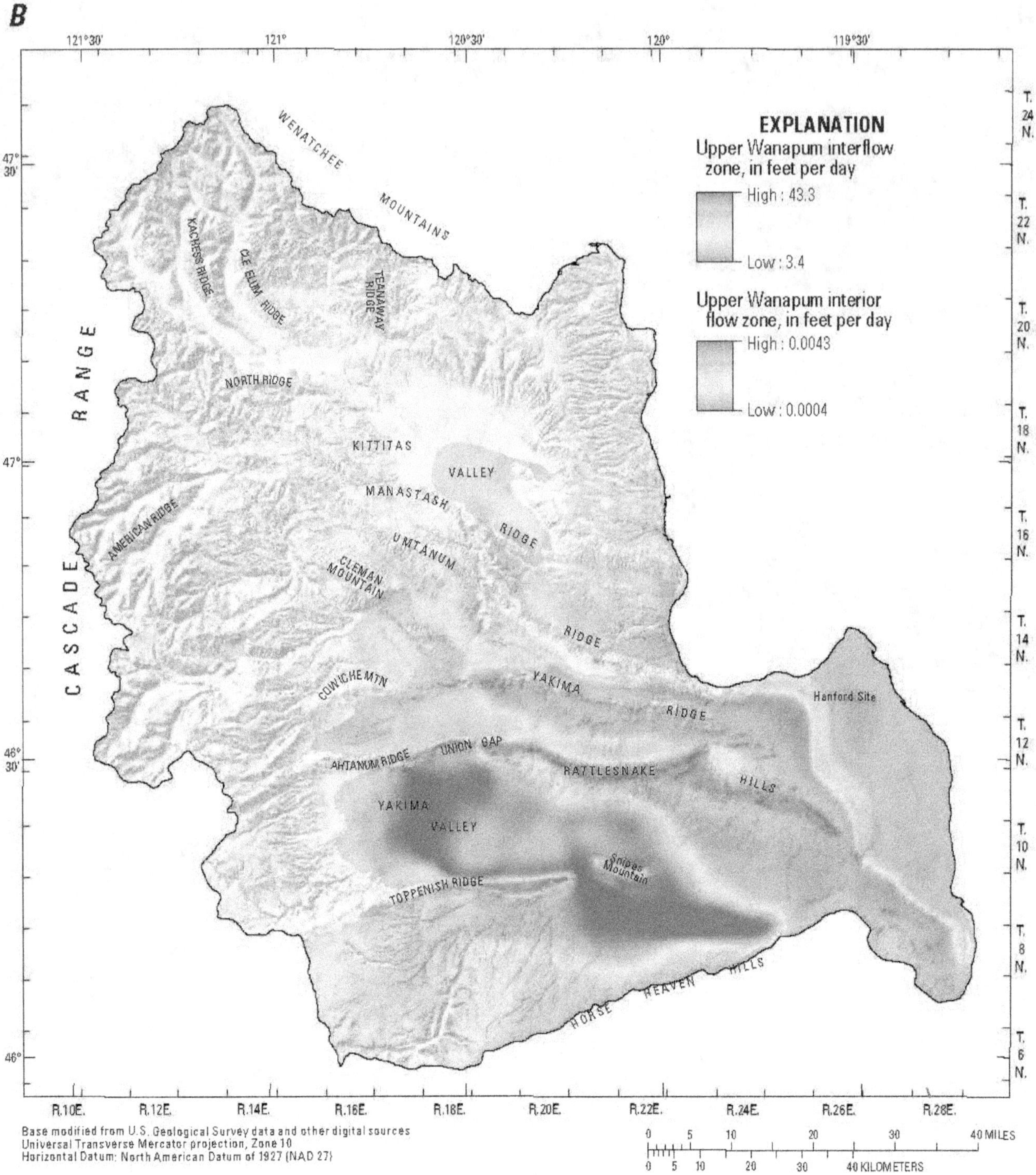
B
121°30'
121°
120°30'
120°
119°30'
47° 30'
47°
46° 30'
46°
T. 24 N.
T. 22 N.
T. 20 N.
T. 18 N.
T. 16 N.
T. 14 N.
T. 12 N.
T. 10 N.
T. 8 N.
T. 6 N.
R.10E.
R.12E.
R.14E.
R.16E.
R.18E.
R.20E.
R.22E.
R.24E.
R.26E.
R.28E.
EXPLANATION
Upper Wanapum interflow zone, in feet per day
High : 43.3
Low : 3.4
Upper Wanapum interior flow zone, in feet per day
High : 0.0043
Low : 0.0004
WENATCHEE MOUNTAINS
KACHESS RIDGE
CLE ELUM RIDGE
TEANAWAY RIDGE
CASCADE RANGE
NORTH RIDGE
KITTITAS VALLEY
MANASTASH RIDGE
AMERICAN RIDGE
UMTANUM RIDGE
CLEMAN MOUNTAIN
COWICHE MTN
YAKIMA RIDGE
Hanford Site
UNION GAP
AHTANUM RIDGE
RATTLESNAKE HILLS
YAKIMA VALLEY
Snipes Mountain
TOPPENISH RIDGE
HORSE HEAVEN HILLS
Base modified from U.S. Geological Survey data and other digital sources
Universal Transverse Mercator projection, Zone 10
Horizontal Datum: North American Datum of 1927 (NAD 27)
0 5 10 20 30 40 MILES
0 5 10 20 30 40 KILOMETERS

Figure 16.—Continued

C

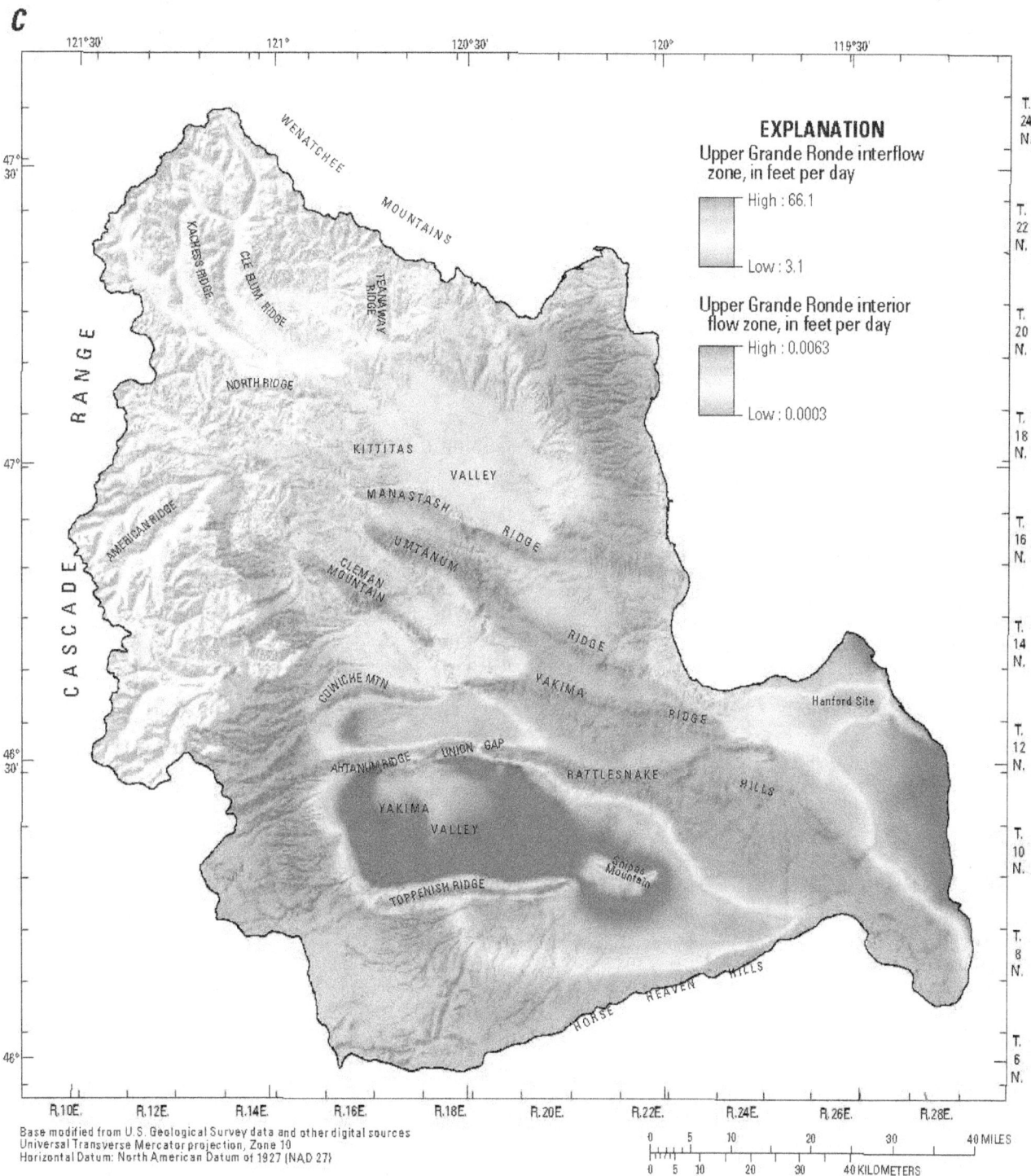

EXPLANATION
Upper Grande Ronde interflow zone, in feet per day

High : 66.1

Low : 3.1

Upper Grande Ronde interior flow zone, in feet per day

High : 0.0063

Low : 0.0003

Base modified from U.S. Geological Survey data and other digital sources
Universal Transverse Mercator projection, Zone 10
Horizontal Datum: North American Datum of 1927 (NAD 27)

Figure 16.—Continued

Vertical Hydraulic Conductivity

Vertical hydraulic conductivity (K_v) values were initially derived from ratios (vertical anisotropy) of the horizontal to the vertical values ($K_h:K_v$). Because of the unknown nature of and local variations in K_v, the large thickness of many of the basin-fill units, and the large range in values presented by other investigators (Vaccaro and others, 2009), anisotropy ratios were regionalized using only two initial anisotropy ratios for the basin-fill units. The ratio for coarse-grained basin-fill units (HGUs 3, 11, 14, 16, 18, and 20) was assumed to be 10:1, and for finer grained units (model HGUs 2, 10, 13, 15, 17, and 19), the ratio was initially assumed to be 100:1. Initial simulations with these values provided reasonable results.

Each basalt unit initially was assigned a uniform vertical anisotropy ratio for the interflow zones and flow interiors. The ratios for the interflow zones of the three mapped HGUs (model HGU s 21, 23, 25, 27, 29, 32, 34, 36, 39, 41, 43, and 45) ranged from 500:1 to 1,000:1. The ratios for the flow interiors (model HGUs 22, 24, 26, 28, 30, 33, 35, 37, 40, 42, 44, and 46) were assigned values that ranged from 0.05:1 to 0.1:1. These lower ratios produced higher values of K_v in the flow interiors than in the interflow zones—which represent the assumption that vertical flow dominates in flow interiors. The Mabton and Vantage Interbeds (model HGUs 31 and 38) were assigned ratios of 10:1 and 1,000:1, respectively.

Storage Properties

Storage coefficient values are known to be highly variable and there is a general lack of information for making reliable areal estimates. However, selected published values for most mapped HGUs were documented in Vaccaro and others (2009), and these values were obtained from aquifer tests and groundwater-modeling studies. Initial constant values for the basin-fill and bedrock units were assigned based on this information. Values for basalt units were spatially varied in proportion to K_h values.

Both unconfined and confined conditions occur within the groundwater system; however, in order to prevent the drying of model cells, and resultant model instability, specific storage values were assigned to cells simulating both unconfined and confined conditions. The specific storage value assigned to the basin-fill units ranged from 0.05 to 0.0001 ft^{-1}. The larger value was for the alluvial aquifers, unit 1 in the Roslyn basin (model HGU 1), and the Pasco gravels (model HGU 16). The thick, coarse-grained unconsolidated unit in the Toppenish basin (unit 2, model HGU 11) was assigned a value of 0.001 ft^{-1}, and the remaining 17 basin-fill units (HGUs 2–10, 12–15, and 17–20) were assigned a specific storage value of 1.0×10^{-4} ft^{-1}.

Values were assigned to the basalt units based on interflow zones and flow interiors. Initial values varied based on the spatial distribution of K_h. For the Saddle Mountains unit, the initial values ranged from 0.0016 to 0.002 ft^{-1} for the interflow zones (model HGUs 21, 23, 25, 27, and 29) and from 1.6×10^{-5} to 2.0×10^{-5} ft^{-1} for the flow interiors (model HGUs 22, 24, 26, 28, and 30). Wanapum interflow zones and flow interiors (model HGUs 32–37) were assigned the same values as those for the Saddle Mountains, and the values have the same range. Storage coefficient values for the Grande Ronde interflow zones and flow interiors ranged from 8.3×10^{-4} to 0.002 ft^{-1} (model HGUs 39, 41, 43, and 45) and from 8.3×10^{-5} to 2.0×10^{-4} ft^{-1} (model HGUs 40, 42, 44, and 46), respectively. The Mabton and Vantage Interbeds (model HGUs 31 and 38) were assigned a uniform value of 0.001 ft^{-1}. The bedrock units (model HGUs 47 and 48) were assigned a uniform value of 1.0×10^{-4} ft^{-1}.

Geologic Structures

Previous work indicates that the CRBG K_h is reduced by one to two orders of magnitude in areas of intense folding and faulting, especially fault-associated anticlines (Hansen and others, 1994; Packard and others, 1996; Reidel and others, 2002). Smaller values along geologic structures may be due to the offsetting of interflow zones through faulting, which produces low-conductivity fault breccia and gouge material at that interface [as described by Stearns (1942) and Newcomb (1965, 1969)], and the reduction of pore space through deposition of secondary minerals. Sharp folding and faulting can cause shearing and fracturing of the basalt flows and create local areas with large permeabilities in joints and fractures, whereas shear faulting can offset the interflow zones and disrupt their hydraulic continuity. Displacement of individual basalt flows along faults, however, locally can enhance vertical movement of water by providing fractured zones across basalt flows that could serve as conduits for vertical groundwater flow, as shown by a methane plume downgradient of a major fault bordering the Hanford Site (Reidel and others, 2002).

The low-permeability geologic structures in the model domain were simulated using the Horizontal-Flow Barrier package (HFB) of Hsieh and Freckleton (1993). Flow barriers comprising 70,139 model cells were grouped into three categories as to their control on the groundwater-flow system: Initial hydraulic characteristic values of the horizontal-flow barrier were 2.0×10^{-6}, 2.7×10^{-5}, and 4.5×10^{-5} 1/d, with the smallest hydraulic characteristic having the largest control (fig. 17). The hydraulic characteristic of the horizontal-flow barrier is the barrier hydraulic conductivity divided by the width of the horizontal-flow barrier along the flow path (leakance).

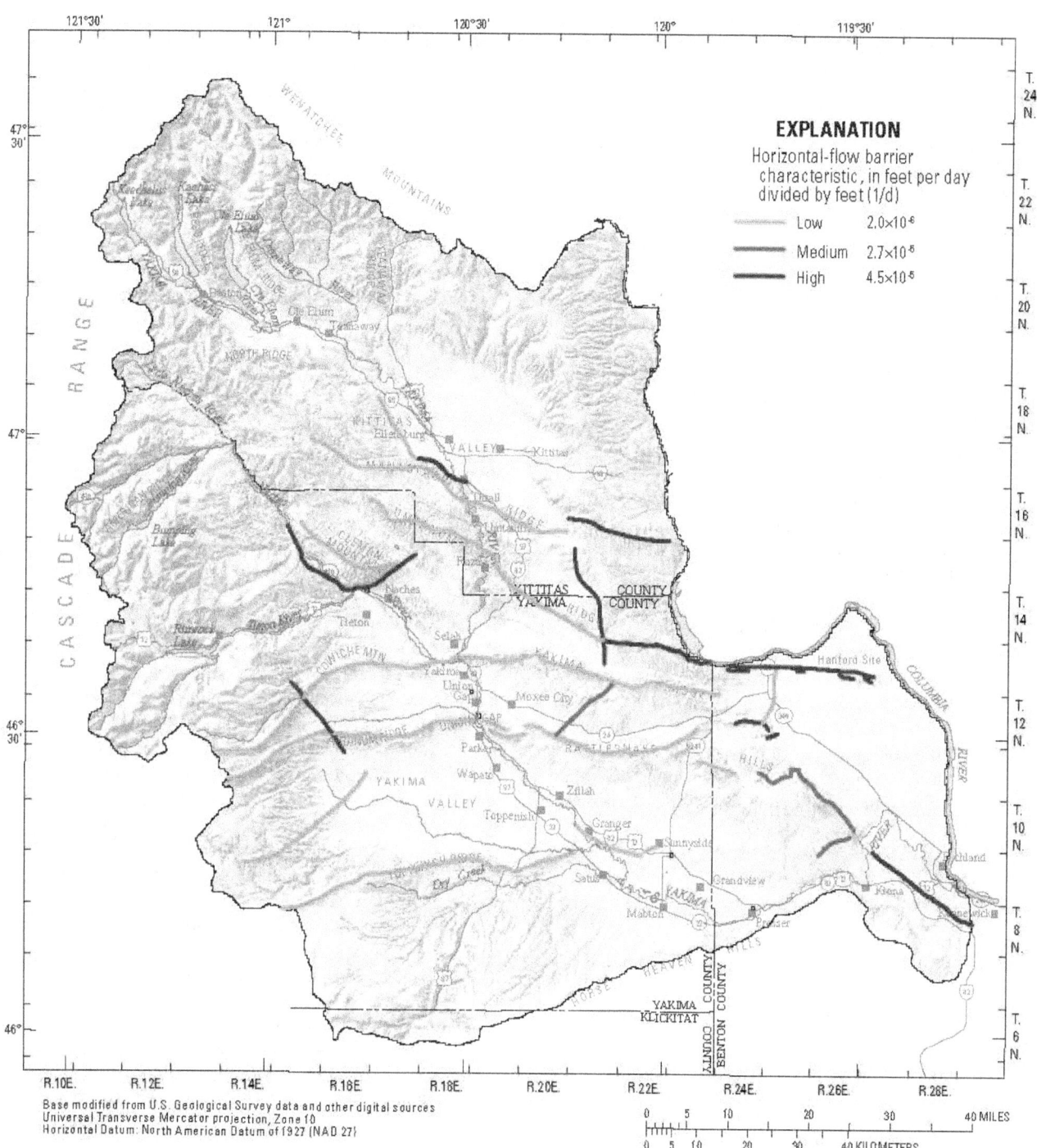

Figure 17. Location of model structure and flow-barrier conductivity, Yakima River basin aquifer system, Washington.

Where buried beneath basin-fill sediments, the structures were assumed to be absent in the overlying sediments. As described previously, there are numerous structures in the basin, ranging from large-scale folds to very minor faults. Two categories of structures were defined for inclusion into the ECM as flow barriers. The first category included the major anticlines such as Umtanum, Toppenish, and Ahtanum Ridges and some smaller anticlines such as Snipes Mountain. The second category was for structures (faults and smaller folds) that are known to affect the groundwater-flow system based on water-level data. Such faults include the Cold Creek fault bordering the northwest part of the Hanford Site and the Bird Canyon Fault near the divide between the Moxee subbasin and the Cold-Dry Creek valleys. Note that Kirk and Mackie (1993) identified the Bird Canyon Fault as the principal hydrologic barrier for the Wanapum and Grande Ronde units, but not for the Saddle Mountains unit, whose east-west barrier was identified as the Hog Ranch-Naneum anticline and the Meyer anticline. To simplify the regional model, the Bird Canyon Fault also was assumed to be the barrier for the Saddle Mountains unit. The importance of geologic structures in the model domain for compartmentalizing the flow system and affecting lateral flow has long been recognized and information on the compartmentalization is summarized in Vaccaro and others (2009). Although synclines are not included as part of the HFB package, they are explicitly accounted for in the spatial distribution of K_h, Hansen and other (1994) and the CPMod derived higher values of K_h in the synclinal basins than along ridges, and this information formed the basis for the spatial distribution of K_h in the basalt units. Descriptions of higher K_h in the synclinal basins were summarized in Vaccaro and others (2009) based on numerous previous studies.

Faults can commonly act as barriers but also can act as preferential flow routes. At this scale, given model cells are 1,000 ft on a side, HFBs were not used to simulate preferential flowpaths.

Model Calibration and Sensitivity

Model calibration is the process in which hydraulic properties (model parameters) are adjusted to obtain a reasonable fit between simulated heads and fluxes and measured data. An integral component of the calibration process is conducting a sensitivity analysis, which provides information on the relative importance of the properties as measured by changes in model fit to measured data. Constructing a complex model with more parameters than the data support may reduce the residuals (differences between measured and simulated values) but does not ensure a more accurate, reliable model (Hill, 1998). Like many groundwater models, the final ECM deviates from the principle of parsimony, but the regionalization of parameters and use of information from previously calibrated groundwater models limits this deviation. Throughout the calibration process, no adjustments were made that conflicted with the general understanding of the aquifer system and previously documented information.

Poorly quantified characteristics of the flow system can be constrained in the calibrated model on the basis of measured hydraulic-head, head-change, and streamflow. Calibration of this model benefited from a particularly extensive set of water-level and streamflow measurements. Streamflow observations (fig. 18, table 5) were used as calibration constraints or targets in the model and predominantly were based on monthly mean streamflow measurements at gaging stations in the basin. Simulated streamflow observations for some of the smaller streams and tributaries were based on gage data from both active and discontinued gages and on simulated values from watershed models (Mastin and Vaccaro, 2002). Water levels measured in wells were used to develop head and head-change observations for model calibration. The USGS NWIS database contained 2,196 wells having 20,279 water-level measurements (fig. 19). Even with this many measurements, model observations are sparse in some areas at some depths, and the paucity of data is most pronounced in the north and northwest part of the model domain, and in the humid uplands (mostly National Forest lands). The distribution of observations by model layers (fig. 20A) shows that the number of head observations decreases with depth in the system, with 48 percent of the observations being in layers 1–4 (64 percent by layer 7); the percentile distribution generally matches the distribution of points of withdrawal by layer. Data for wells with water-level measurements over multiple years were available for analyzing temporal variations. The open intervals of the wells were considered in determining the model layers associated with head and head-change observations.

Temporally, water-level data existed throughout the simulation period and was distributed fairly evenly (fig. 20B). Of the 20,279 total measurements used in the calibration, 8 percent were from water years 1960 to 1969, 24 percent from 1970 to 1979, 27 percent from 1980 to 1989, 13 percent from 1990 to 1999, and 28 percent from 2000 to 2001. The greater number of water-level measurements during 2000 and 2001 than in previous years was in part due to the intensive field effort associated with the study, and the larger percent during 1980–89 is due to the many measurements taken as part of developing the CPMod.

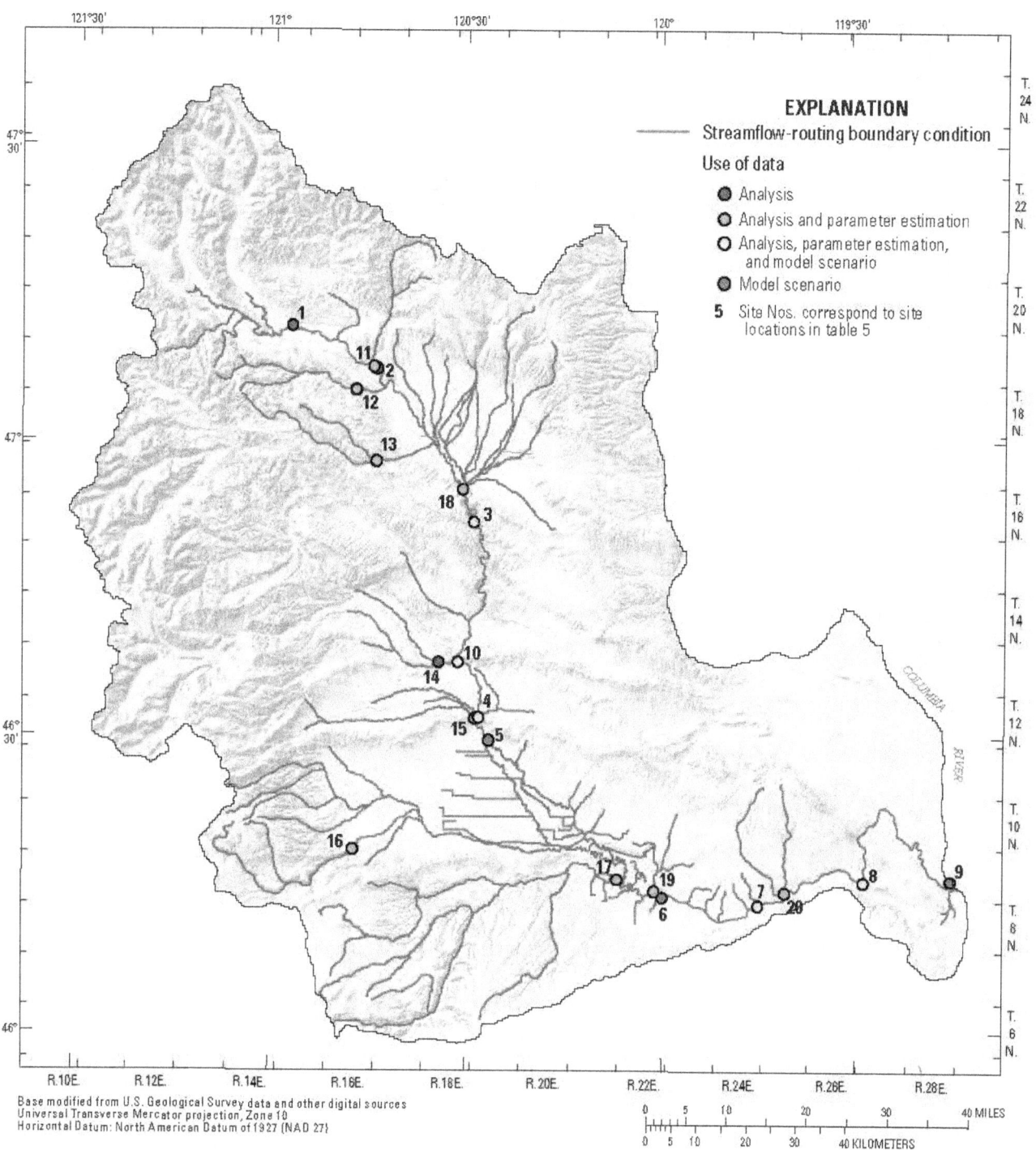

Figure 18. Location and use of model streamflow observations, Yakima River basin aquifer system, Washington.

Table 5.   Streamflow measurement sites used for analysis, parameter estimation, and (or) scenarios, Yakima River basin aquifer system, Washington.

[Site locations are shown in figures 1, 3, and 18. **Source of data:** E, estimated; GS, U.S. Geological Survey; M, miscellaneous measurements; R, Reclamation; RS, Roza-Sunnyside Irrigation districts; SIM, simulated; W, watershed models; YN, Yakama Nation. **Use of data:** AN, analysis; PE, parameter estimation; SCEN, scenario]

Site name	Name used in report	Source of data	Use of data	Site No. in figure 18
	Upstream prescribed flow	W		
RIVERS				
Yakima River at Cle Elum	Cle Elum	R	AN	1
Yakima River at Horlick	Horlick	E, R	AN	2
Yakima River at Umtanum	Umtanum	GS	AN, PE, SCEN	3
Yakima River above Ahtanum Creek, at Union Gap	Union Gap	E, GS	AN, PE, SCEN	4
Yakima River near Parker	Parker	GS, R	SCEN	5
Yakima River at Mabton	Mabton	GS	SCEN	6
Yakima River at Prosser	Prosser	R	AN, PE, SCEN	7
Yakima River at Kiona	Kiona	GS	AN, PE, SCEN	8
Yakima River at Richland	Richland	SIM	SCEN	9
Naches River at North Yakima	Naches	R, E	AN, PE, SCEN	10
CREEKS				
Swauk Creek	Swauk Creek	W	AN, PE	11
Taneum Creek	Taneum Creek		AN, PE	12
Manastash Creek	Manastash Creek	W	AN, PE	13
Wenas Creek	Wenas Creek	W		
Cowiche Creek	Cowiche Creek	W	AN	14
Wide Hollow Creek	Wide Hollow Creek	W		
Ahtanum Creek	Ahtanum Creek	GS	AN, PE	15
Toppenish Creek	Toppenish Creek	W	AN, PE	16
Satus Creek	Satus Creek	W	AN, PE	17
DRAINS - CREEKS				
Bedrock Drains	Bedrock Drains	R, GS		
Wilson-Cherry Creeks	Wilson-Cherry Creeks	R	AN, PE	18
Granger Drain	Granger Drain	RS, GS		
Marion Drain	Marion Drain	M, YN		
Sulphur Creek	Sulphur Creek	R	AN, PE	19
Spring-Snipes Creek	Spring-Snipes Creek	E	AN, PE	20

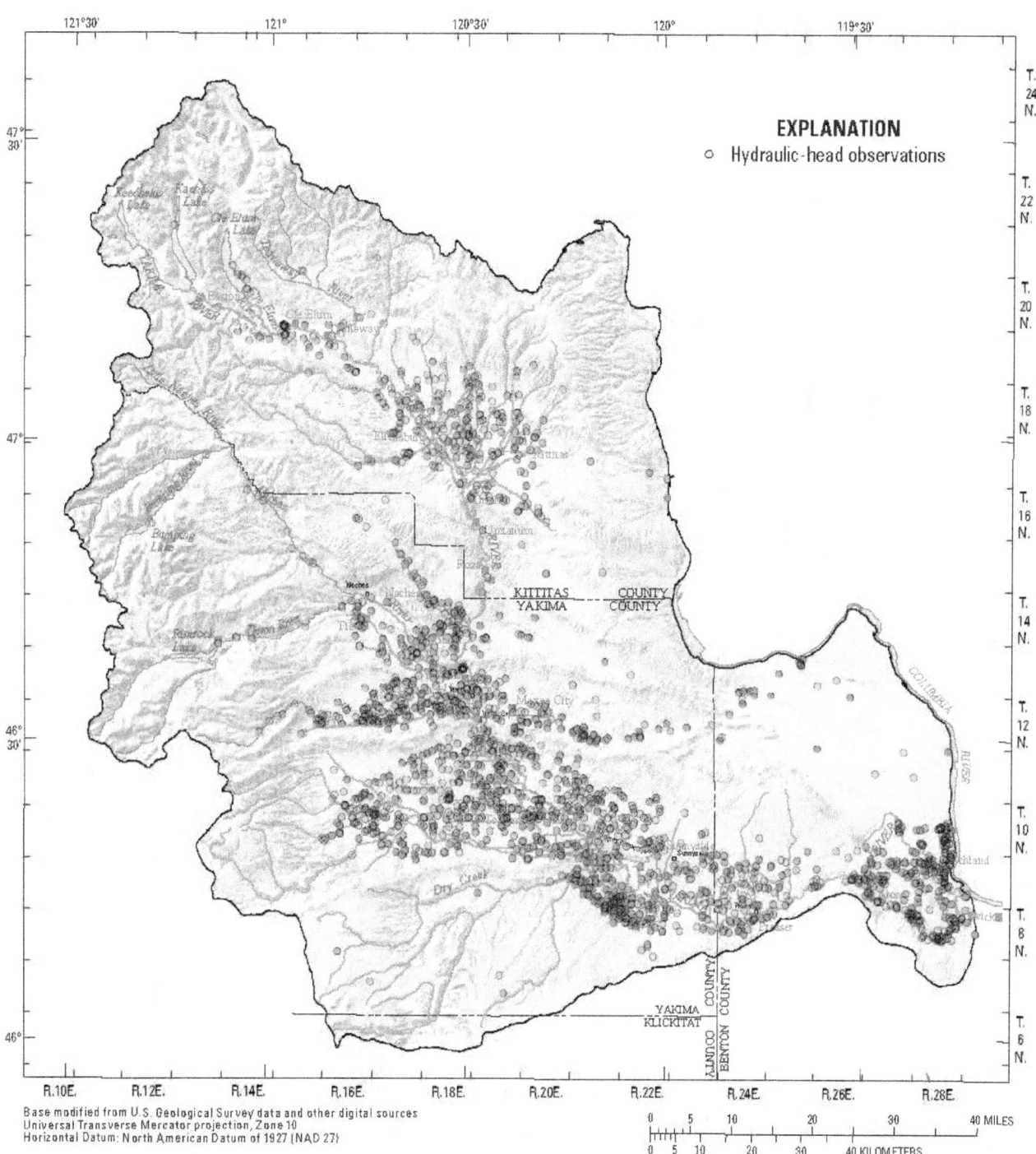

Figure 19.   Location of model hydraulic-head observations, Yakima River basin aquifer system, Washington.

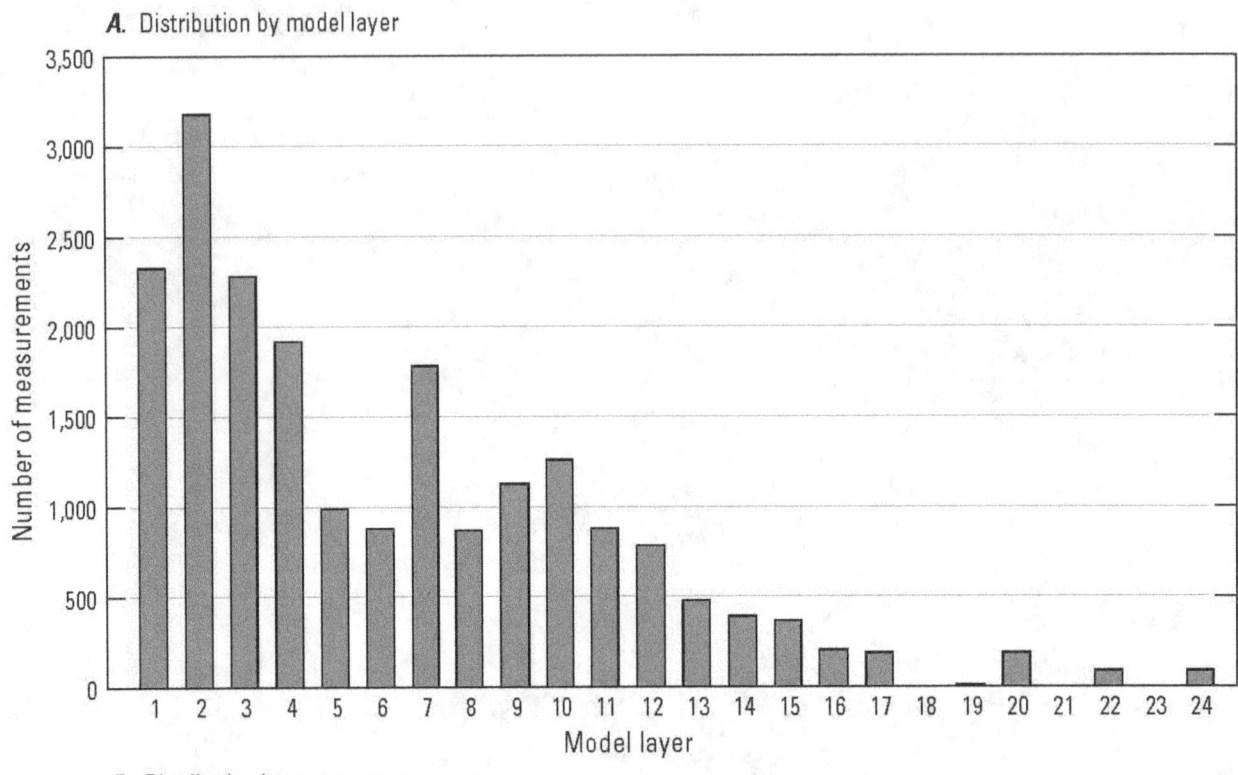

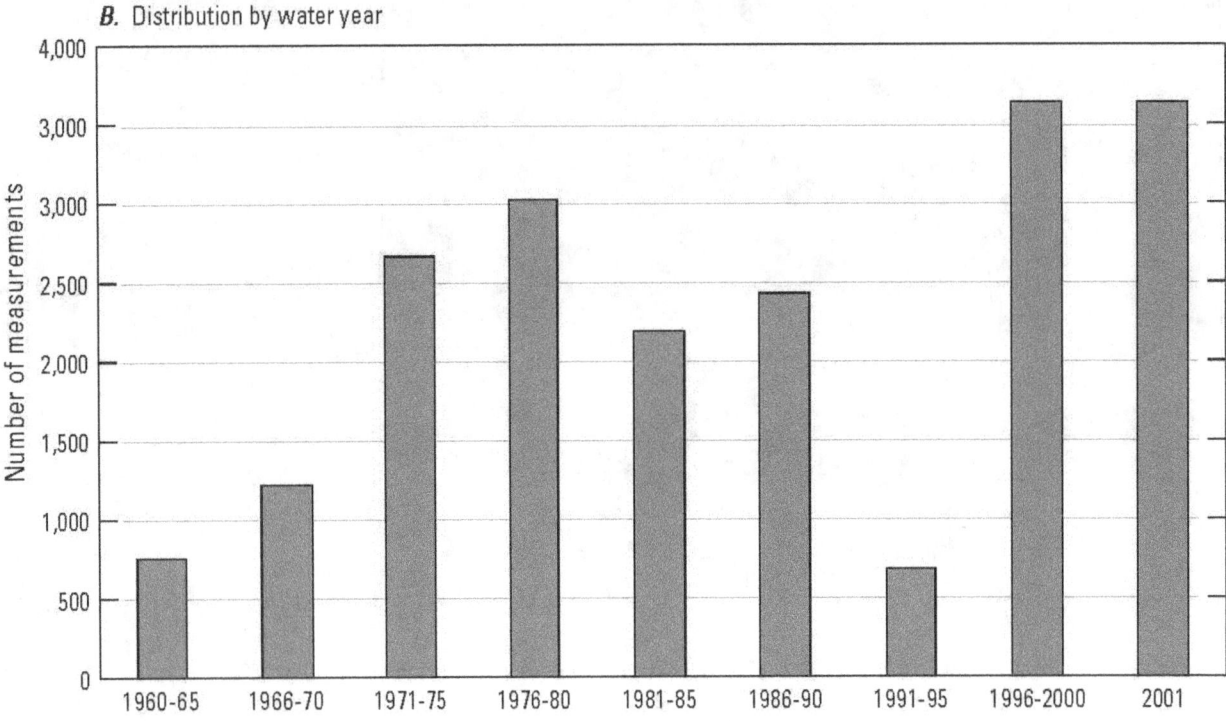

Figure 20. Distribution of model hydraulic-head observations by model layer and water years, Yakima River basin aquifer system, Washington.

Calibration Approach

The model was calibrated using the iterative parameter estimation software package PEST (Doherty, 2010). PEST uses a nonlinear least-squares regression to find the set of parameter values that minimizes the weighted sum-of-squared-errors objective function.

Observations and, therefore, residuals are weighted to allow a meaningful comparison of measurements with different units (weighted residuals are dimensionless) and to reduce the influence of measurements with large errors or uncertainty. The initial observation weight was defined as the inverse of the variance of the measurement error, but modifications to this weighting scheme were used to account for discrepancies in data density between observation groups.

The parameters adjusted during calibration included K_h, K_h:K_v (vertical anisotropy), specific storage, stream conductances, and the hydraulic conductivity of structural barriers. Drain conductances were adjusted but only in terms of producing a total groundwater discharge to the drains that approximated the prescribed inflows to the SFR2 package. Structural barriers were divided into three groups with each barrier assigned to one group with a common hydraulic characteristic, thus representing 40 geologic structures with only three parameters.

A total of 151 model parameters were specified, but these were reduced to 97 estimated parameters by holding constant the ratio of K_h, K_h:K_v, and specific storage for each interflow zone and flow interior within a basalt group. Thus, the 10 model HGUs, 21–30, defined for the Saddle Mountain basalts were represented in the model calibration by two multipliers for each of three parameters, K_h, K_h:K_v, and specific storage. For each parameter, one multiplier represented all five Saddle Mountain interflow zones and one multiplier represented all five flow interiors, reducing the 30 initial Saddle Mountain parameters to 6 estimated parameters. This process of parameter simplification preserved the relative changes in parameter values assigned on the basis of depth corrections as described in section, "Horizontal Hydraulic Conductivity" while still allowing interflow zones and flow interiors to be separately estimated for each basalt group. The same process was applied to the six Wanapum model HGUs, 32–37 (18 parameters reduced to 6 estimated parameters) and the eight Grande Ronde model HGUs, 39–46 (24 parameters reduced to 6 estimated parameters).

Model calibration was first conducted using a trial-and-error process to ensure that model predictions were in reasonable agreement with measured trends in groundwater levels and streamflow variations. Subsequent automated calibrations of the 97 estimated model parameters using parallel PEST were conducted using 20,279 head observations spanning the entire 42-year simulation period and 7,056 monthly streamflow observations at 14 gage sites. Calibrations using both heads and flows were conducted with observation weights adjusted to ensure equal contribution by the two groups to the model objective function, in accordance with USGS guidelines (Doherty and Hunt, 2010).

Sensitivity Analysis

Sensitivity analysis is used to assess the effects of different conceptual models (different model designs and parameter values) on the simulated heads and discharges, and to develop useful nonlinear regressions (Hill, 1998; Hill and Tiedeman, 2003). The ability to estimate a parameter value using nonlinear regression is a function of the sensitivity of simulated values to changes in the parameter value. Parameter sensitivity reflects the amount of information about a parameter provided by the observation data. Generally speaking, if a parameter has a high sensitivity, observation data exist to effectively estimate the value. If the parameter has low sensitivity, observation data are not sufficient to estimate the parameter value and changing the parameter value will have little effect on the sum of squared errors. Parameter sensitivities change with changes in parameter values during calibration, and are expected to be lowest overall for the well-calibrated model.

Parameter Correlation Coefficient

Parameter correlation coefficients indicate if two or more parameters can be uniquely estimated (optimized) by nonlinear regression. They are calculated as the covariance between two parameters, divided by the product of the standard deviations. Correlation coefficients range from -1.0 to 1.0 and absolute values larger than 0.95 may indicate a possible high degree of correlation. If extreme parameter correlation exists, the correlation coefficient will be close to 1.0 or -1.0. A low degree of correlation implies that the action of one parameter is independent of the action of another parameter with regard to the value of the simulated output. The implications of non-unique parameter values are explained by Poeter and Hill (1998).

Observations Used in Model Calibration

Water Levels, Water-Level Changes, and Associated Errors

The hydraulic-head data used for calibration consisted of 20,279 water-level measurements from 2,196 wells made between 1960 and 2001. Latitudes and longitudes for the well locations were determined by two methods—for the 14,615 wells with groundwater levels measured generally prior to 2000, wells were located by well drillers' or as part of previous investigations that located numerous wells on 1:24,000 topographic quadrangles. The accuracy of legacy well location and water-level data may not be fully known, but careful examination of the data found that most sites were within an acceptable margin of error for this study. Questionable data were removed from the dataset and not used in the calibration. For water levels measured during the 2000 field effort, a Global Positioning System (GPS) with a horizontal accuracy of one-half second (about 50 ft) was used to determine latitude and longitude at wells visited during the current study. Depth to water (water level) was measured in most wells using a calibrated electric tape or graduated steel tape, both with accuracy to 0.01 ft. At deeper high-capacity wells, an airline measurement was taken using the airline at the well, a calibrated pressure gage, and a tank containing compressed air. Altitude of the land surface at each site was interpolated from the 10-m DEM, with an accuracy of ±8 ft (one-half the contour interval). Few water-level measurements in wells completed in bedrock were used in the calibration. Information from bedrock wells was used, however, to help understand the connection between the basalt units and the bedrock units.

The observation weights initially were assigned using the methods suggested by Hill (1998). Hydraulic-head measurement errors were limited by the accuracy of locating the well on a topographic map and the accuracy of the topographic map used to determine land-surface altitude at wells whose locations were not measured using a GPS. For wells located using a GPS, altitude accuracy is limited by the accuracy of the DEM used to estimate the altitude. The accuracy of the water-level measurements (assumed to be 0.01 ft) was subtracted from a land-surface altitude with an accuracy of ±8 ft to estimate a measurement error. For calibration purposes, all water-level observations were assigned equal relative weights.

Model calibrations conducted using observations of different types require a weighting scheme that adequately represents the contribution to total model error of observations made in different measurement units. In this case, the 20,279 water-level measurements contributed a model error measured in feet while the 3,232 streamflow observations contributed a model error measured in cubic feet per second. Several weighting schemes designed to balance the contribution of these two observation types

based on measurement error were tried, but the prevalence of water levels over mean monthly flows consistently produced model calibrations that did not capture streamflow variability adequately for the purposes for which the model was designed. To redress this imbalance, weights for each class of observations were proportionally scaled such that water-level observations and streamflow observations each made roughly equivalent contributions to total model error. Due to the large number of water-level measurements used, all head observations were assigned equal relative weights and all variability in observation weighting was applied to the streamflow observations.

Model observations for water levels were assigned to model layers based on the well depth. However, water levels measured in the field are measurements of hydraulic head throughout the open-well interval, and thus, the water levels represent an average head throughout the screened interval. Wells used in the calibration of the groundwater model had as many as nine separate screen intervals spanning an average of 72 ft. Seventy-five wells had open intervals greater than 500 ft. Most wells that are open to multiple layers are basalt unit wells. For wells open to more than one model layer, simulated heads from the lowest model layer to which the well was open were compared to measured heads in the well. Large screened intervals are likely to contribute to larger residual head errors during the calibration process. This method provided a consistent basis for comparisons but may have resulted in large residuals for cases where a simulated head from a single model layer was compared to a measured head from an open interval that spanned more than one mapped HGU. The multi-node well package (MNW2) for MODFLOW (Konikow and others, 2009) was created to represent wells open to multiple intervals, but greatly increases model run times and instability. The approach presented here does not use MNW2 but was considered sufficient for a regional model of this scale.

Further complicating measured water levels as model observations is the issue of intraborehole flow. Intraborehole flow allows mixing of water from one layer with water from another layer. The water level in such a well would represent an average head for all screened intervals. Erick Burns (U.S. Geological Survey, written commun., 2010) found that aquifer cross-connection by wells is the dominant process driving the measured head declines in the CRBG in the Mosier basin, Oregon. The Mosier basin represents an extreme case of folding, faulting, and compartmentalization, making the intraborehole flow a major cause in the very large head declines. Burns was unable to simulate the head loss without simulating the effects of intraborehole flow. However, in the central Columbia Plateau, where there are numerous large-capacity, deep, uncased wells, the CPMod simulated the areas and magnitude of groundwater-level declines and the regional flow system with a high degree of accuracy (Hansen and others, 1994). Intraborehole flow was not directly simulated in the ECM, thus introducing possible structural error into the model. MNW2 can simulate intraborehole flow in wells that are connected to more than one node of the finite-difference grid and may have reduced the error, but at considerable cost in model computational efficiency and numerical stability.

Streamflow Observations and Errors

Streamflow observations used to constrain the model calibration came from several sources. Measured streamflow at USGS and Reclamation streamflow-gaging stations and streamflow simulated using watershed models were all used as observations in model calibration. Synoptic streamflow measurements made along the Yakima River and its tributaries to quantify the groundwater discharge to, or recharge from, the surface-water system were used to estimate the ability of the ECM to simulate gains and losses along the river system. Information for streamflow observations used in either model calibration, testing model goodness of fit, analysis of flow system, or model applications are shown in table 5; shortened names of sites used herein are presented in the table.

Streamflow measurements used to calibrate the ECM also were weighted by using a proportional inverse of the measured flow, thus assuming equal relative confidence intervals for all flow measurements regardless of flow magnitude. Because streamflow measurements vary from a few hundred to tens of thousands of cubic feet per second, smaller streamflow observations without this weighting scheme would contribute little to model error and the model calibration would be biased toward peak flows largely ignoring low flows. The proportional inverse weighting scheme was applied exclusively and uniformly to streamflow measurements, and thus assumes that both large and small flow measurements carry the same inherent percent measurement error. After proportional inverse weighting, all streamflow measurements were then uniformly scaled up by an experimentally determined multiplier until the initial contribution to total model error was equivalent for the sum of water-level observations and the sum of streamflow observations. This weighting scheme violates the methods and guidelines explained in Hill (1998), but was considered necessary to ensure that streamflow observations were meaningfully included in model calibration and to reduce the influence of large relative errors during peak streamflow.

Model-Calibrated Hydraulic Properties and Parameter Uncertainty

Final model parameter values are shown in figure 21 and table 6. The specific weighting scheme resulted in parameter confidence intervals that ranged from about 85 to 115 percent of the parameter value. The confidence intervals for hydraulic conductivity and stream conductivity parameters spanned a reasonable range, indicating sufficient information existed to estimate the parameter value.

As anticipated, calibrated hydraulic conductivities were consistently lower for all basalt flow interiors than for the corresponding basalt interflow zones. Upper bedrock layers were determined to be more conductive than lower bedrock layers. The conductivities of basin fill sediments were within the ranges of previously published values and displayed internally consistent relative values both within each subbasin and between subbasins.

Vertical anisotropy ratios for the three basalt groups showed values comparable to basin fill sediments for basalt interflow zones, and lower values for flow interiors. These relative values are consistent with the observed physical characteristics of basalt flows, including high horizontal conductivity in the brecciated and rubbly interflow zones and high vertical conductivity in flow interiors attributed to the columnar jointing common throughout the Columbia Plateau.

Model parameter sensitivities shown in figure 22 show the relative importance of model parameters to head observations only and streamflow observations only. The parameters most sensitive to hydraulic-head observations are those for the hydraulic conductivities of basalt interflow zones and flow interiors for each of the three major basalt units, as well as several of the basin-fill sediments from which the largest volumes of groundwater pumpage occurred and water-level observations were made. The most sensitive specific storage parameters are those for the Mabton Interbed (model HGU 31) and the Wanapum interflow zones and flow interiors (model HGUs 32–37), although storage for the flow interiors of the Grande Ronde (model HGUs 40, 42, 44, and 46) also is notably significant. Parameter sensitivities to streamflow observations had a much different pattern. Composite-scaled sensitivities for specific storage showed little variability, with all normalized sensitivities within an order of magnitude from the largest (Mabton Interbed, model HGU 31).

Combined sensitivities for head and flow observations are less variable than for either head or flow observations alone due to the differing impacts of each observation type on the various listed parameters. Total observation sensitivities were greatest for the Grande Ronde interflow zones (model HGUs 39, 41, 43, and 45) and upper bedrock horizontal conductivities (model HGU 47), which is consistent with their regional extent and common contribution to both head and flow observations. Combined specific storage sensitivities were highest for the regional Mabton Interbed (model HGU 31), which separates the Saddle Mountain and Wanapum basalt groups, and for basin-fill model HGU 3, the basal gravels of the Roslyn subbasin.

If two parameters are highly correlated, they cannot be uniquely estimated. The largest positive and negative parameter correlation coefficients for the calibrated model presented here were 0.4233 (special storage in model HGU 31 to K_v in model HGU 47) and -0.5368 (K_h in model HGU 8 to K_h in model HGU 9), suggesting that all parameters can be uniquely estimated.

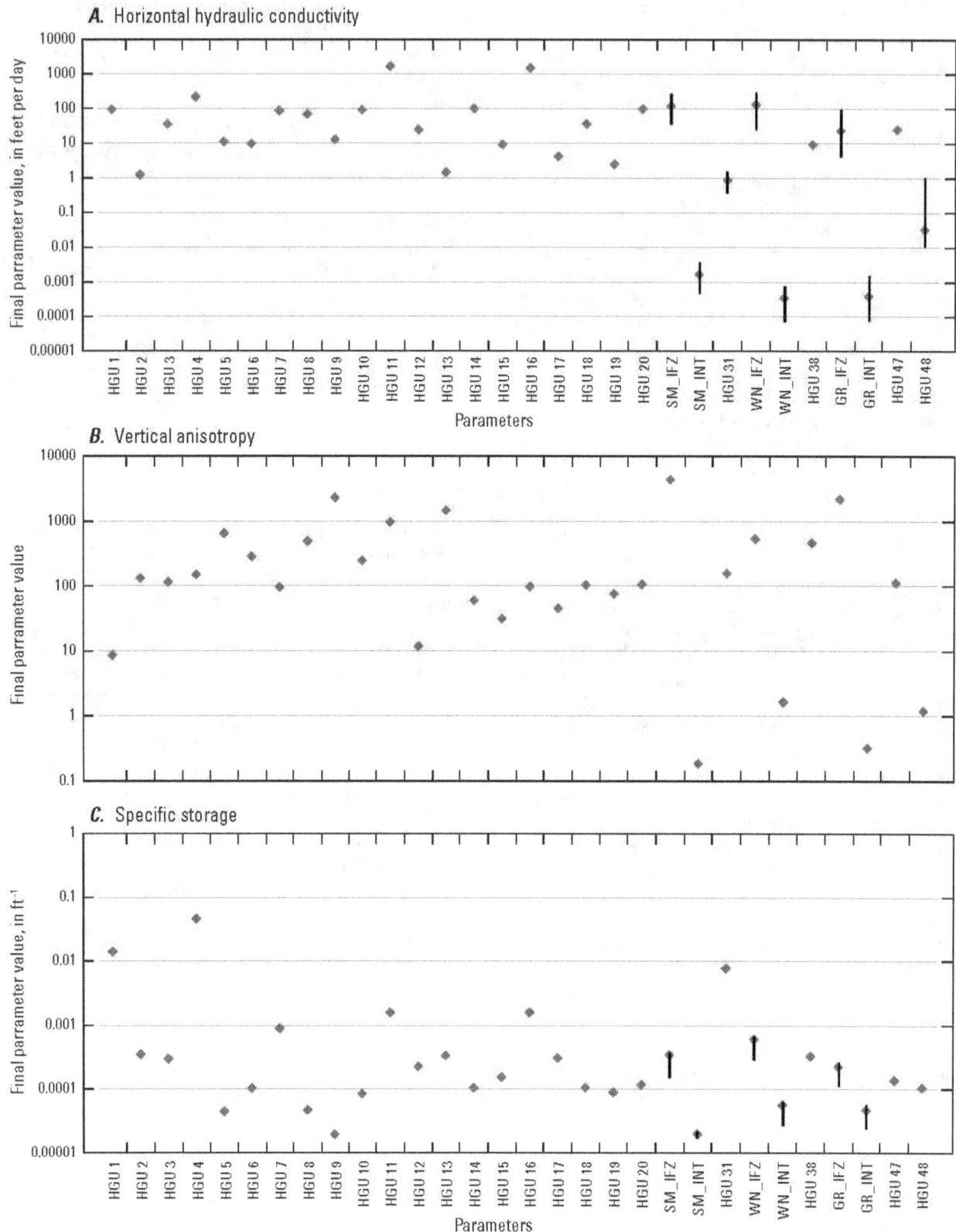

Figure 21.   Final calibrated model parameter values and ranges and means of parameter values, Yakima River basin aquifer system, Washington.

Table 6.   Final calibrated model parameters for the Yakima River basin aquifer system, Washington.

[**Abbreviations**: HGU, hydrogeologic unit; SM, Saddle Mountains unit; WN, Wanapum unit, GR, Grande Ronde; IFZ, interflow zone; INT, flow interior; SFR, Stream-Flow Routing package; DRN, Drain package; HFB, Hydrologic-Barrier Flow; ft/d, foot per day; ft²/d; square foot per day]

Hydrogeolgic unit	Horizontal hydraulic conductivity (ft/d)				Vertical anisotropy	Specific storage (ft⁻¹)			
	Minimum	Mean	Maximum	Constant		Minimum	Mean	Maximum	Constant
HGU 1				93.37	8.66				1.39E-02
HGU 2				1.24	133.32				3.45E-04
HGU 3				35.73	115.24				2.92E-04
HGU 4				215.71	150.51				4.61E-02
HGU 5				11.13	644.11				4.40E-05
HGU 6				9.67	282.53				1.02E-04
HGU 7				86.44	96.05				8.84E-04
HGU 8				69.21	485.42				4.62E-05
HGU 9				12.89	2,302.48				1.92E-05
HGU 10				91.74	246.19				8.33E-05
HGU 11				1,672.30	959.54				1.56E-03
HGU 12				24.34	11.78				2.22E-04
HGU 13				1.46	1,449.37				3.28E-04
HGU 14				99.92	60.19				1.03E-04
HGU 15				9.25	31.58				1.52E-04
HGU 16				1,490.37	98.53				1.57E-03
HGU 17				4.19	45.83				3.02E-04
HGU 18				36.07	104.19				1.04E-04
HGU 19				2.51	76.11				8.77E-05
HGU 20				98.24	106.43				1.15E-04
SM_IFZ	35.78	119.15	261.43		4,336.83	1.51E-04	3.39E-04	3.73E-04	
SM_INT	4.91E-04	1.66E-03	3.56E-03		0.19	1.68E-05	1.96E-05	2.15E-05	
HGU31	0.36	0.88	1.57		157.39				7.70E-03
WN_IFZ	25.67	129.94	278.32		530.33	2.83E-04	5.99E-04	6.76E-04	
WN_INT	6.92E-05	3.51E-04	7.49E-04		1.66	2.68E-05	5.60E-05	6.33E-05	
HGU38				9.31	461.85				3.22E-04
GR_IFZ	4.28	22.99	90.97		2,177.27	1.11E-04	2.19E-04	2.65E-04	
GR_INT	7.61E-05	3.95E-04	1.54E-03		0.32	2.43E-05	4.64E-05	5.76E-05	
HGU 47				24.88	112.90				1.35E-04
HGU 48	0.01	0.03	1.09		1.21				1.03E-04

Streamflow-routing cells   Conductivity (ft/d)
Yakima River   0.45
Tributaries   0.75

Drain cells   Conductance (ft²/d)
Headwater streams   323,166.00

Hydraulic-flow barrier   Hydraulic characteristic (1/d)
Low   5.18E-06
Medium   3.59E-05
High   4.20E-04

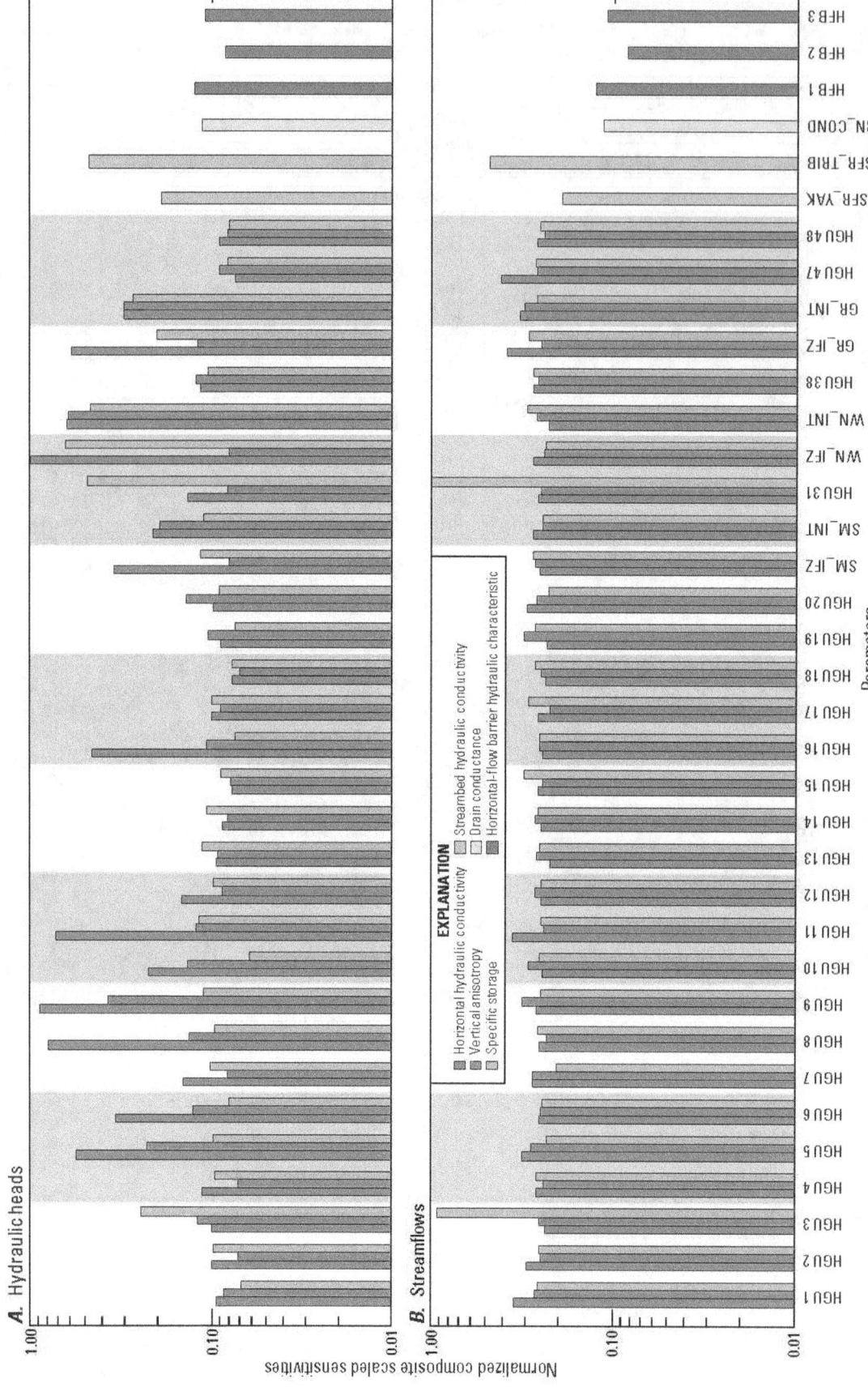

Figure 22. Normalized composite scaled sensitivities of final calibrated model parameters to hydraulic-head observations and streamflow observations, Yakima River basin aquifer system, Washington.

Statistical Measures of Model Fit

The measure of model fit can be represented with many statistical and graphical methods. One measure of model fit is based on the difference between simulated and measured heads and flows, or residuals. The overall magnitude of the residuals is considered, but the distribution of those residuals, both statistically and spatially, can be equally important. The magnitude of residuals can initially point to gross errors in the model, the data (measured quantity), or how the measured quantity is simulated (Hill, 1998). A complete discussion of the statistical measures discussed in this section is found in Hill (1998).

A commonly used indicator of the overall magnitude of the weighted residual is the calculated error variance. The calculated error variance is the weighted sum-of-squared errors divided by the number of observations, minus the number of parameters. The square root of the calculated

error variance is called the standard error of regression. Smaller values of both terms indicate a closer fit between the simulated and measured values. The calculated error variance and standard error of regression are dimensionless and therefore not intuitively informative about goodness of fit. Model calibration reduced the calculated error variance from 1.523×10^6 to 1.318×10^4 and the standard error of weighted residuals from 684.7 to 114.8. For the hydraulic heads alone, the standard error of regression is 124 ft.

A useful graphical analysis is a simple plot of all (hydraulic heads and streamflow) weighted measured values as a function of all weighted simulated values. The residuals should be normally distributed along a line with a slope equal to 1.0 and a y-intercept of zero. The weighted observations versus weighted simulated values generally fall along a straight line with a slope of 1.02 and a y-intercept of 88.8 (fig. 23).

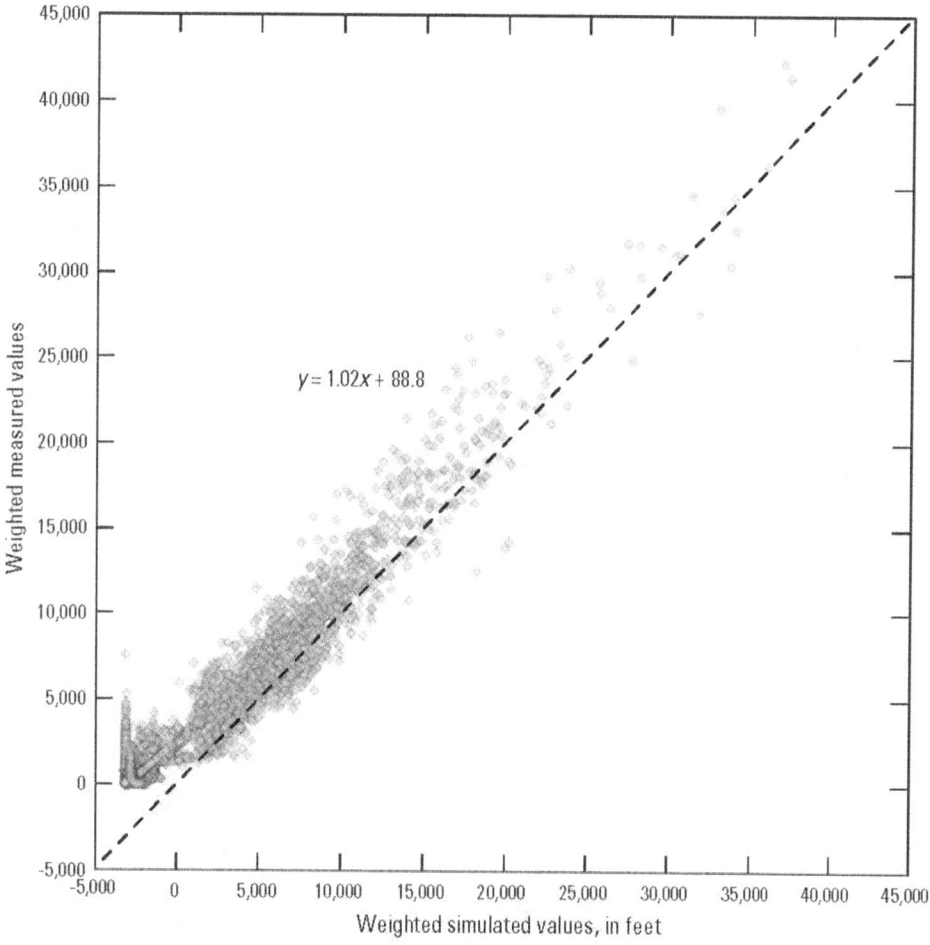

Figure 23. Weighted measured values as a function of weighted simulated values, Yakima River basin aquifer system, Washington.

Transient Calibration Model Fit

A graphical and descriptive comparison of simulated and measured values provides a clear insight to the model fit and complements the statistical measures of model fit described above. Such a comparison indicates how the model replicates the flow system, including water levels and streamflow. Although the error variance for the ECM model is well within an acceptable limit, it is important to determine that the model simulates the regional groundwater-flow system (directions and amounts of flow), as the ability to simulate the flow system may be unrelated to the measures described above.

Comparison of Simulated and Measured Hydraulic Heads

A traditional and intuitive assessment of model calibration is a simple plot of measured hydraulic heads as a function of simulated hydraulic heads (fig. 24). At 2,196 well measurement points, the mean and median difference between 20,179 simulated and measured hydraulic heads is -49 and -13 ft, respectively. The residuals for the 42-year simulation period show that 39 percent of the simulated heads exceeded measured heads with a median residual value of 20 ft, and 61 percent were less than measured heads with a median residual value of -49 ft. The root-mean-square (RMS) error of the difference between simulated and measured hydraulic heads in the observation wells, divided by the total difference in water levels in the groundwater system (Anderson and Woessner, 1992, p. 241), also had to be less than 10 percent to be acceptable (Drost and others, 1999). The calibrated model produces an RMS error divided by the total difference in water levels of 4 percent.

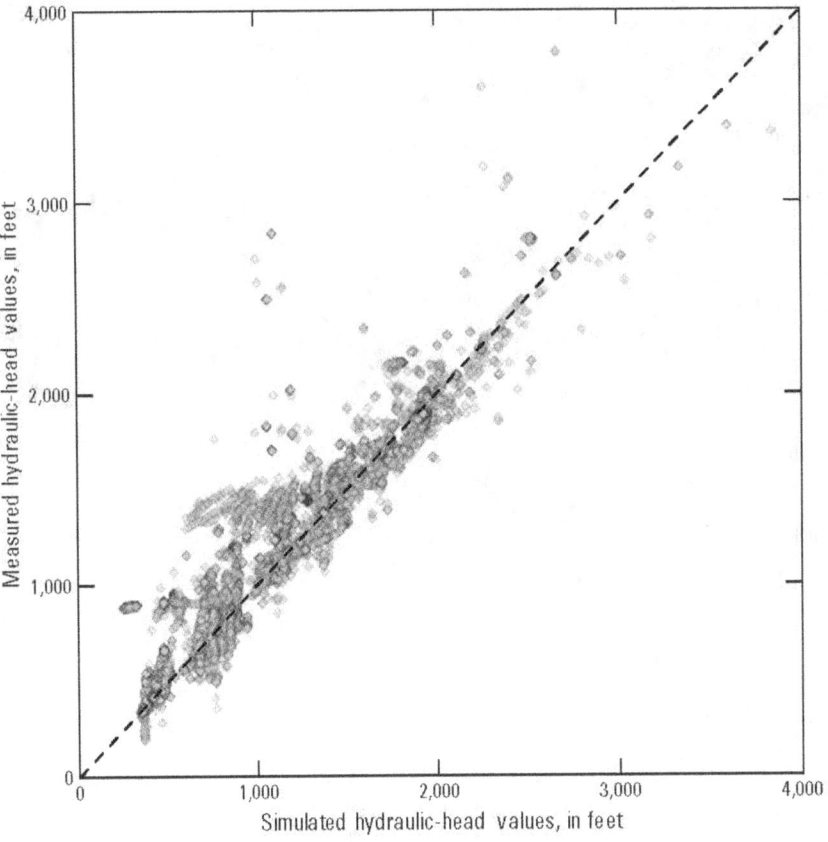

Figure 24.   Measured hydraulic heads as a function of simulated hydraulic heads, Yakima River basin aquifer system, Washington.

The residuals for basin-fill and basalt units indicate a reasonable fit but the spatial distribution of the residuals shows definite patterns of bias (fig. 25). For example, lower heads were simulated in the northeastern part of the model domain in the Hog Ranch-Naneum anticline area where large vertical downward gradients occur. Measured water levels in this area may be representative of perched conditions but there are not sufficient data to verify this. In contrast, calculated heads were higher than the observed along the eastern part of the south slope of Rattlesnake Hills near the boundary of Roza Irrigation District.

Most clusters of largest residuals represent a single well with many water-level measurements. Reasons for the large residuals could be explained by the presence of perched water conditions, a model observation assigned to the wrong model layer, wells screened in multiple HGUs, or not enough recharge specified in the model. K_h is very low in the areas with large residuals and no reasonable parameter adjustment during calibration could raise water levels. Hydraulic characteristic values of the horizontal-flow barriers also were decreased with no improvement in simulated water levels.

For simulated hydraulic heads to be acceptable, the distribution of heads and the patterns of flow also should approximate the generalized water-level distributions and flow patterns mapped as part of this study (Vaccaro and others, 2009). Although a bias exists toward measured hydraulic heads exceeding simulated hydraulic heads, simulated groundwater-flow patterns generally match the mapped patterns (water-level contours have similar shape and thus, flow directions are similar). For example, the simulated depth to water (fig. 26) shows a reasonable match to the mapped contours of Vaccaro and others (2009) with differences in the shape of the contours appearing to be more related to the ECM capturing parts of the regional flow system that were not able to be mapped in detail due to sparse data, and (or) calculating smoother head distributions in some areas such as in the upland areas to the northwest in the Kittitas structural basin. For the latter, simulated heads in the uplands are less detailed than the mapped water levels. Mapped water-level contours were based on sparse data. The simulated overall surficial flow system (fig. 27A) displays complexity that would be expected in such a large-scale, complex regional flow system. Lateral gradients in the uplands range from about 175 to 800 ft/mi and they are less than the landscape gradients—measured groundwater levels show groundwater gradients are less than the slope of the land surface in upland areas. The largest gradients occur in steep canyon areas that are controlled by perennial streams, recharge, and hydraulic characteristics. In the structural basins, the simulated horizontal gradients are nearly identical to those mapped and reported by Vaccaro and others (2009). Regardless of whether simulated heads are higher or lower than measured heads in some areas, the capturing of lateral gradients and flow directions further indicates that the simulated flow system is a reasonable representation of the actual flow system, and thus, the ECM can be used to assess potential effects on the flow system (direction and magnitude of groundwater flow) due to variations in recharge (natural and human-induced) and pumping. For the structural basins, simulated and mapped flow systems generally are similar. Differences generally occur where the mapped water-level contours represent the very shallow system, such as in parts of the Ahtanum subbasin.

The simulated water-level contours for the Saddle Mountains and Wanapum units (figs. 27B and 27C) also display complexity that is similar to the mapped contours of Vaccaro and others (2009). The compartmentalization of the flow system due to geologic structure is clearly defined, especially for the Saddle Mountains unit. As described by Vaccaro and others (2009), this compartmentalization is an important aspect of the flow system as pumping effects generally will be confined to the hydrogeologic units and streamflow within each compartment. The ECM captures the mapped basalt hydraulic-head gradient from steep to shallow where a basalt unit meets the basin-fill units; where the basalt meets the sediments also is an important component of the flow system. The ability to simulate the compartmentalization of the system, lateral gradients, and flow directions further indicates that the ECM captures the major components of the regional flow system.

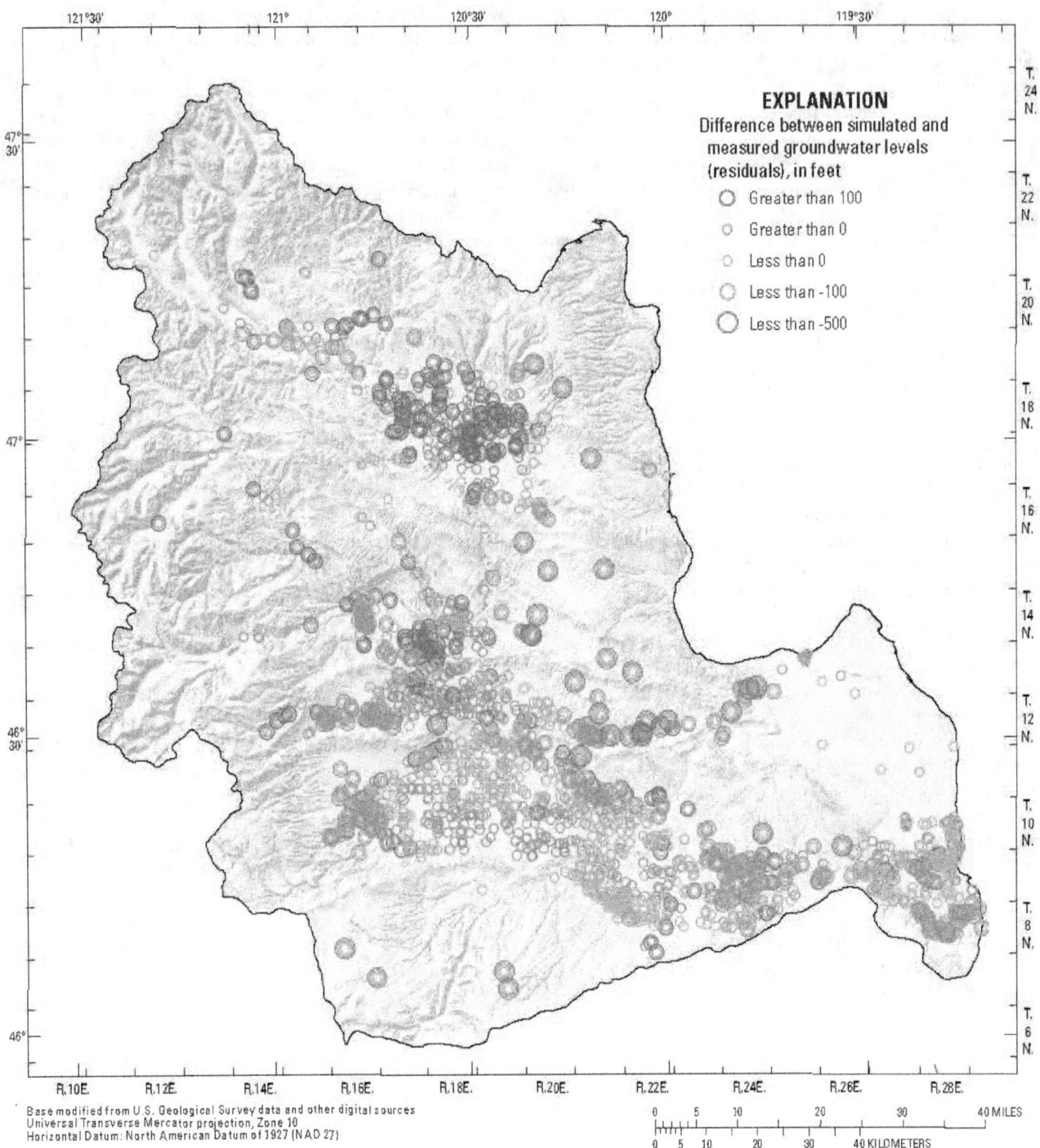

Figure 25.   Differences between simulated and measured groundwater levels (residuals), Yakima River basin aquifer system, Washington.

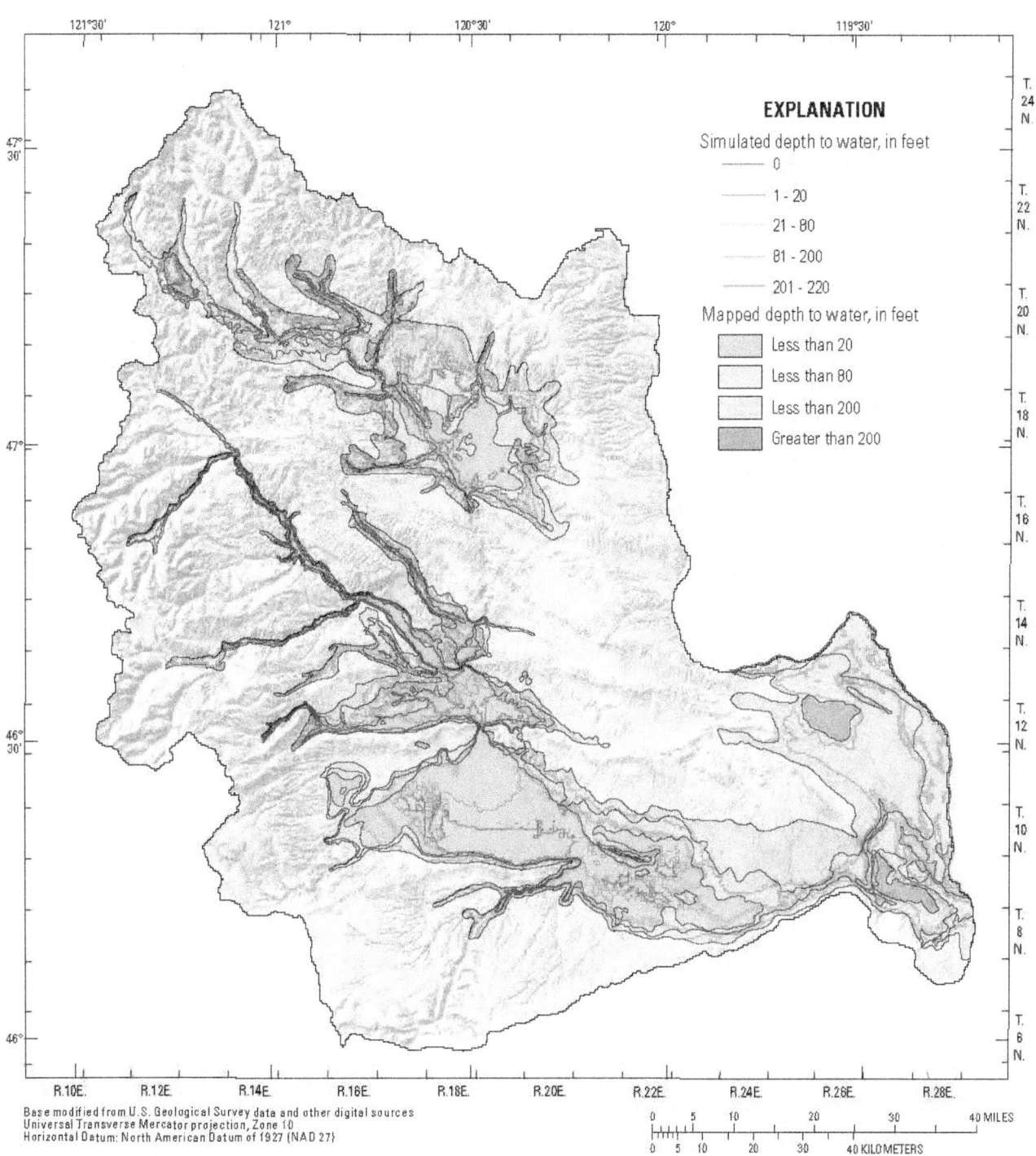

Figure 26.   Simulated and mapped depth to water, Yakima River basin aquifer system, Washington.

A

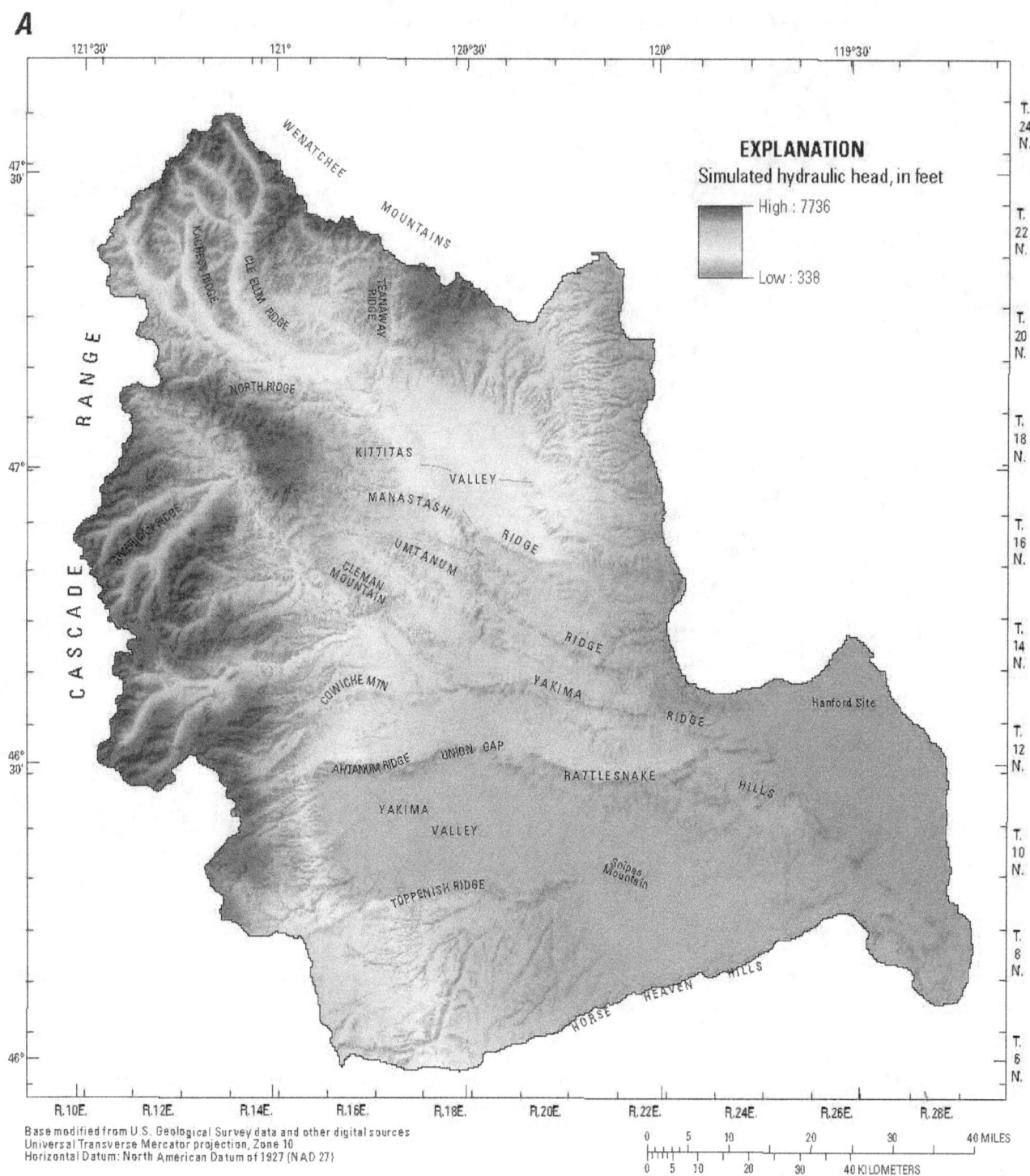

Figure 27.   Simulated hydraulic-head for surficial model layer (*A*), Saddle Mountains unit (*B*), and Wanapum unit, (*C*), Yakima River basin aquifer system, Washington.

B

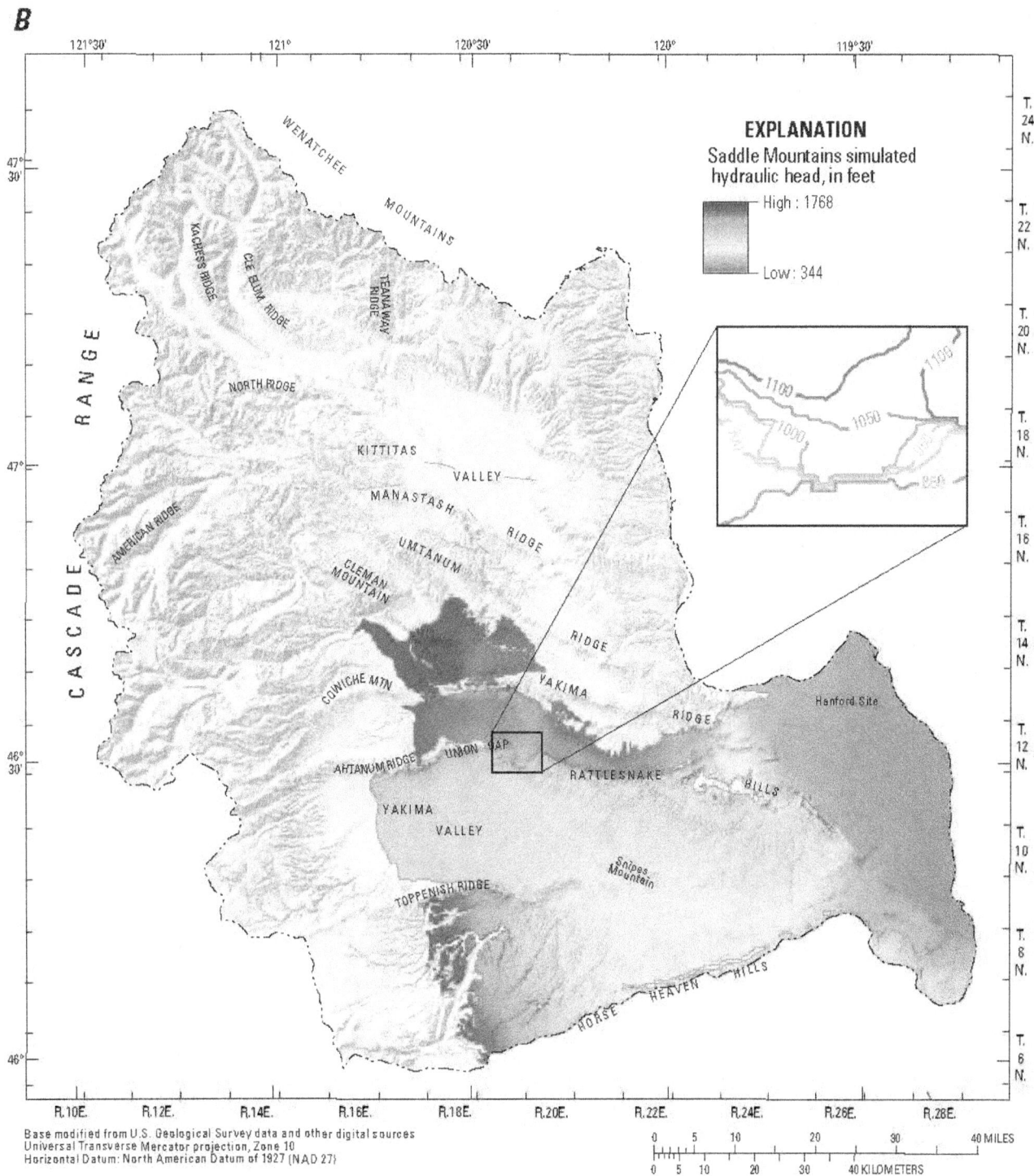

Base modified from U.S. Geological Survey data and other digital sources
Universal Transverse Mercator projection, Zone 10
Horizontal Datum: North American Datum of 1927 (NAD 27)

Figure 27.—Continued

C

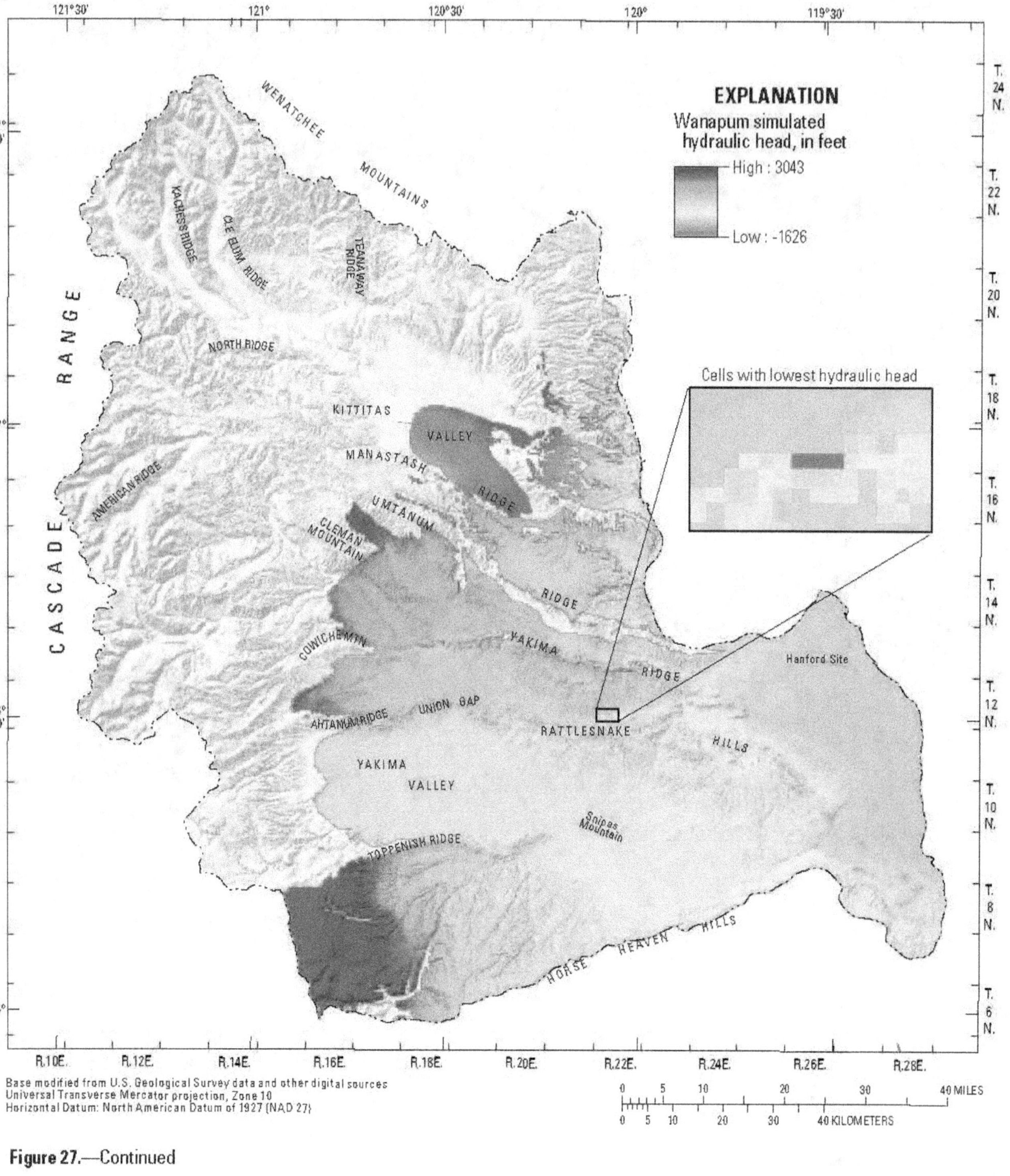

EXPLANATION
Wanapum simulated hydraulic head, in feet
High : 3043
Low : -1626
Cells with lowest hydraulic head
121°30'
121°
120°30'
120°
119°30'
47° 30'
47°
46° 30'
46°
T. 24 N.
T. 22 N.
T. 20 N.
T. 18 N.
T. 16 N.
T. 14 N.
T. 12 N.
T. 10 N.
T. 8 N.
T. 6 N.
R.10E.
R.12E.
R.14E.
R.16E.
R.18E.
R.20E.
R.22E.
R.24E.
R.26E.
R.28E.
WENATCHEE MOUNTAINS
CASCADE RANGE
KACHESS RIDGE
CLE ELUM RIDGE
TEANAWAY RIDGE
NORTH RIDGE
KITTITAS VALLEY
MANASTASH RIDGE
AMERICAN RIDGE
UMTANUM RIDGE
CLEMAN MOUNTAIN
COWICHE MTN
YAKIMA RIDGE
Hanford Site
UNION GAP
AHTANUM RIDGE
RATTLESNAKE HILLS
YAKIMA VALLEY
Snipes Mountain
TOPPENISH RIDGE
HORSE HEAVEN HILLS
Base modified from U.S. Geological Survey data and other digital sources
Universal Transverse Mercator projection, Zone 10
Horizontal Datum: North American Datum of 1927 (NAD 27)
0 5 10 20 30 40 MILES
0 5 10 20 30 40 KILOMETERS

Figure 27.—Continued

Comparison of Simulated and Measured or Estimated Streamflow

The ECM simulates streamflow over 1,616-mi of stream reaches (8,533 model cells). The simulation of streamflow includes routing, calculations of streamflow gains/losses, and accounting for diversions/returns. Thus, simulated streamflow discussed herein explicitly includes the uncertainty associated with diversion rates. Excluding the specified inflows, simulated values do not include a direct surface-water runoff component. A comparison of simulated and measured or watershed-derived streamflow at selected sites in the study area provides additional information on the reliability of the ECM. Streamflow gains and losses also are an important component of the simulated flow-system water budget, especially related to the total-basin water mass balance under regulated conditions. Simulated stream leakage accounts for about 3 percent of the 42-year cumulative water budget for total simulated flows into the aquifer system (streamflow losses) and 16 percent of the total simulated flows out of the aquifer system (streamflow gains).

For the Yakima River, simulated and measured streamflow generally display a close correspondence (fig. 28). The percent differences between mean annual flows for the sites shown on figure 18 for the 42-year period range from less than 1 percent to about 9 percent for the Yakima River sites and 5 percent for the Naches River. For all Yakima River sites and the Naches River, 58 percent of the simulated annual values were within 10 percent of the measured values, and 63 percent were within 15 percent. Of the 346 simulated annual values for the 9 sites, 31 percent were underpredicted (simulated less than measured), and 88 percent were within 25 percent of measured values. Of the overpredicted (simulated greater than measured) annual values, 32 percent of these values were overpredicted by more than 15 percent. Overprediction generally occurs in low flow years, especially from Cle Elum to Union Gap. Note that during low-flow years, an overprediction of 200 ft³/s at Parker could equate to a 24-percent error, whereas a 500 ft³/s error for a high-flow year could equate to a less than 10-percent error. At Parker where annual generally values were slightly underpredicted, 43 percent of the simulated values were within 300 ft³/s. These annual differences indicate that there is a bias towards underprediction, but overpredicted values tend to have a poorer fit. Annual differences between measured and simulated values ranged from less than 1 percent at all Yakima River sites to 96 percent at Prosser in 1994, with the exception of 1977, an extremely low-flow drought year that the ECM did not reproduce. Twenty-nine percent of the 304 simulated annual values were within 5 percent of the measured values.

Of the 4,152 monthly differences for the Yakima and Naches River sites, 54 percent of the monthly simulated values were between 85 and 115 percent of measured values, and 74 percent were between 75 and 125 percent. The median difference was 3 percent. Overall, 15 percent of the monthly values were underpredicted by more than 25 percent and 11 percent were overpredicted by more than 25 percent. As for the monthly values, overprediction generally occurs during the low-flow periods. For example, at Horlick, the measured flow was 590 ft³/s in January 2001 and the simulated value was 900 ft³/s (a difference of 310 ft³/s), whereas the measured flow in August 2001 was 3,000 ft³/s and the simulated flow was 3,400 ft³/s (a difference of 400 ft³/s). Considering the complex system with numerous diversions and returns, the ability to capture the shape of the hydrographs and the overall streamflow values indicates that the simulated mainstem streamflow over most of the 42-year period for most of the sites is within expected error bounds for streamflow simulated with a groundwater model.

The two upper basin sites (Cle Elum and Horlick, fig. 28 and table 5) are upstream areas with the least influence of pumping and diversions; the hydrograph match is good, because most of the streamflow is controlled by specified inflows in the SFR2 package. In some years (for example, 1994–96), there are distinct differences in the hydrographs between Cle Elum and Horlick and the ECM is able to simulate these variations over 21 river miles, if not the absolute measured streamflow. The simulated low flows (which generally began after 'flip-flop' in mid-September) are larger than the measured flows. The larger low flows are likely due to (1) underpredicting tributary flow that results in increased water in the groundwater system that discharges over a longer period of time, and (2) excluding simulation of some laterals/drains in the SFR2 package—including the complete lateral/drain system for all irrigation districts was beyond the scope of this study. Peak flows also are underpredicted because (1) they are supported by surface-water runoff (overland flow) that is not part of the recharge component, (2) underprediction of tributary peak flows, and (3) the 1-month stress period used in the model.

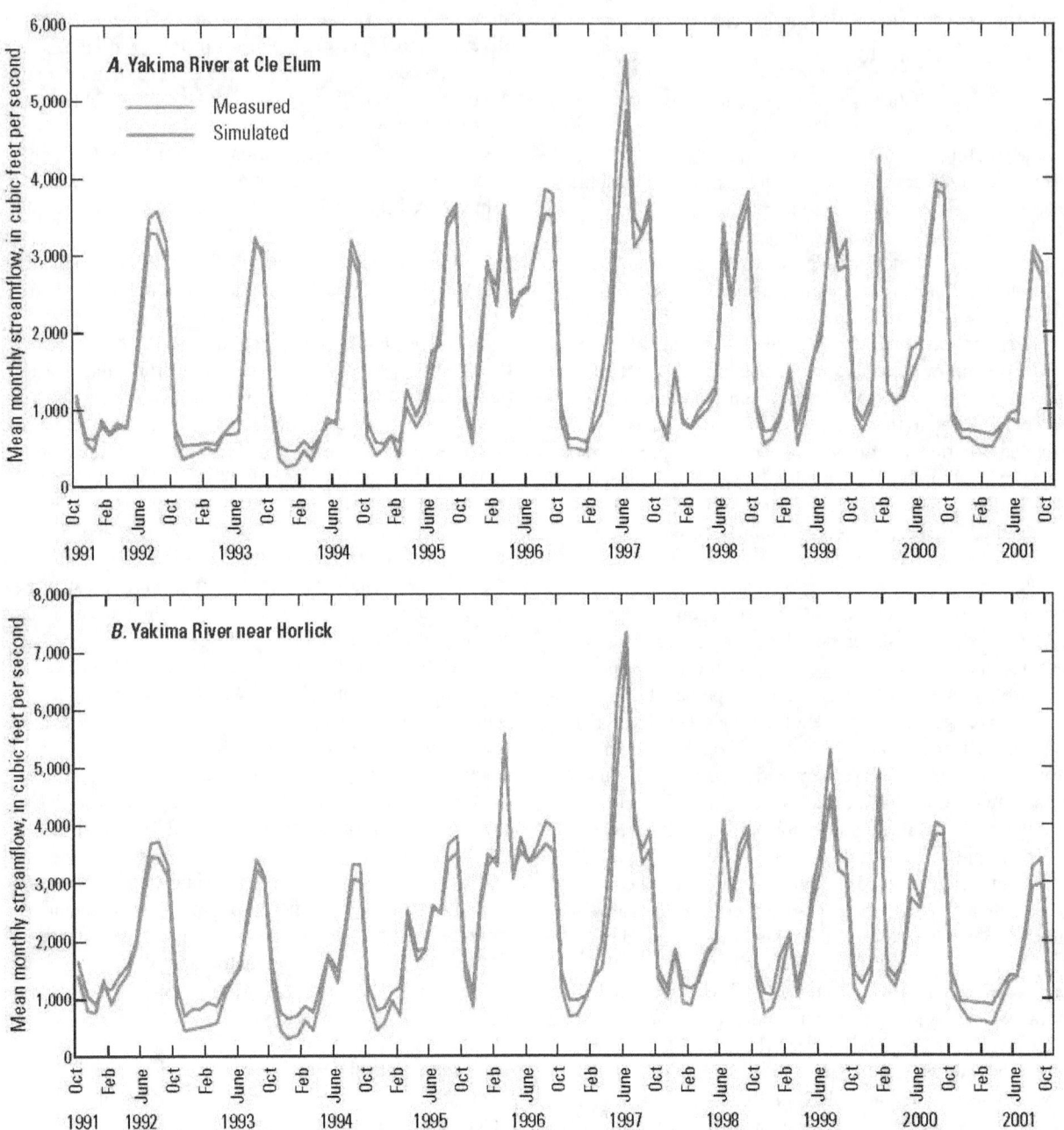

Figure 28.   Simulated and measured mean monthly streamflow of the Yakima River at Cle Elum, Yakima River near Horlick, Yakima River at Umtanum, Naches River near mouth, Yakima River at Union Gap, Yakima River near Parker, Yakima.

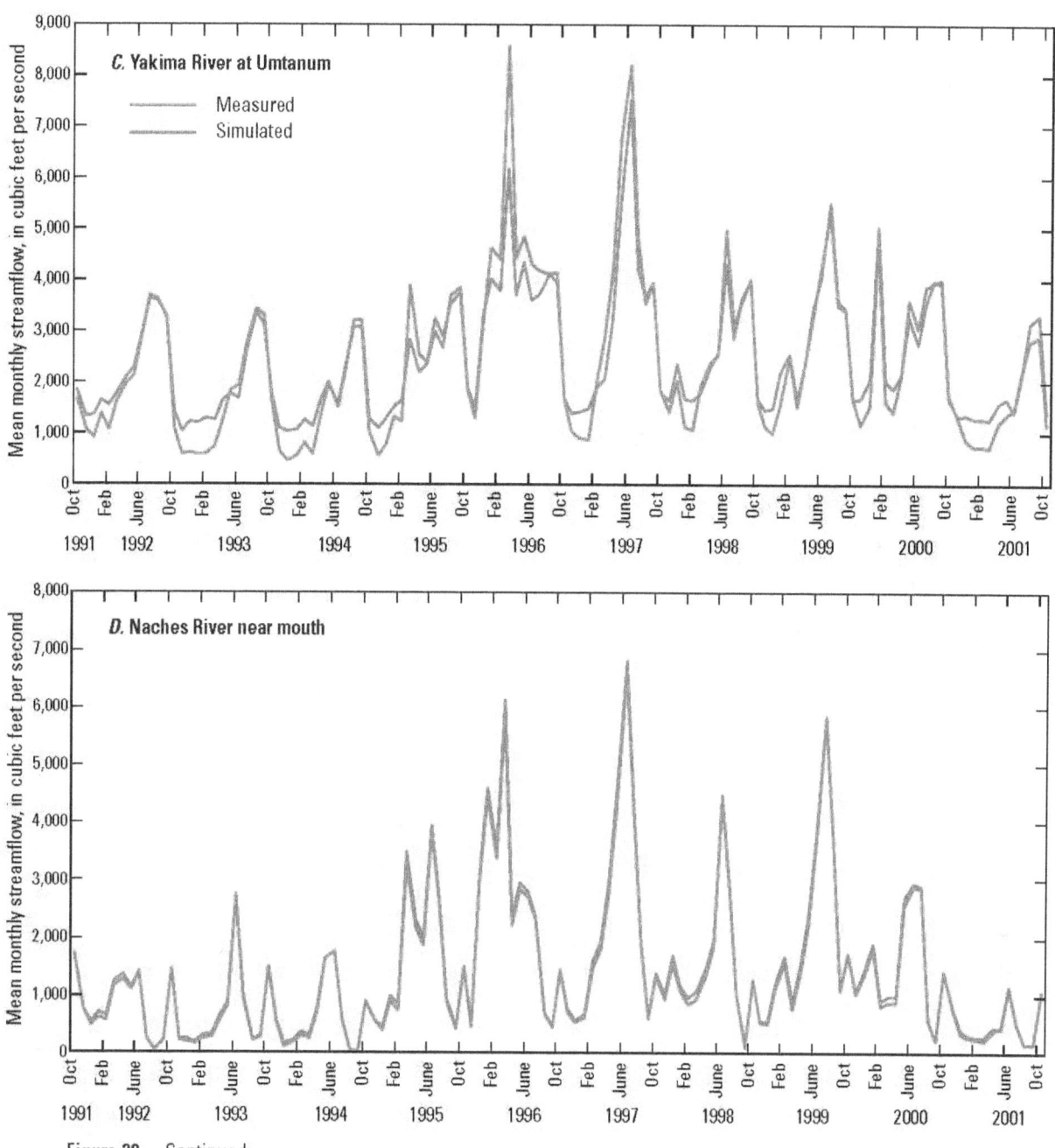

Figure 28.—Continued

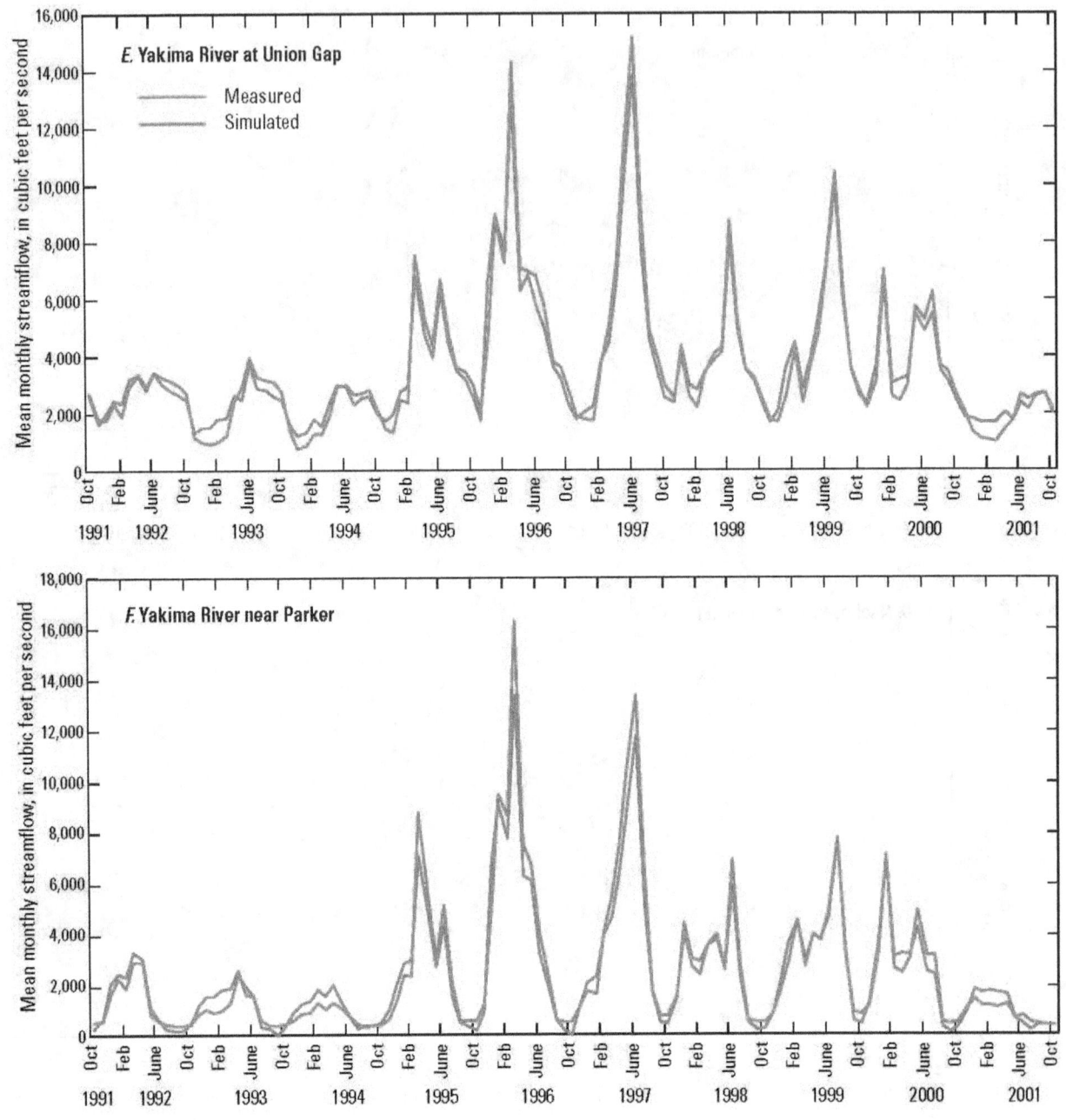

Figure 28.—Continued

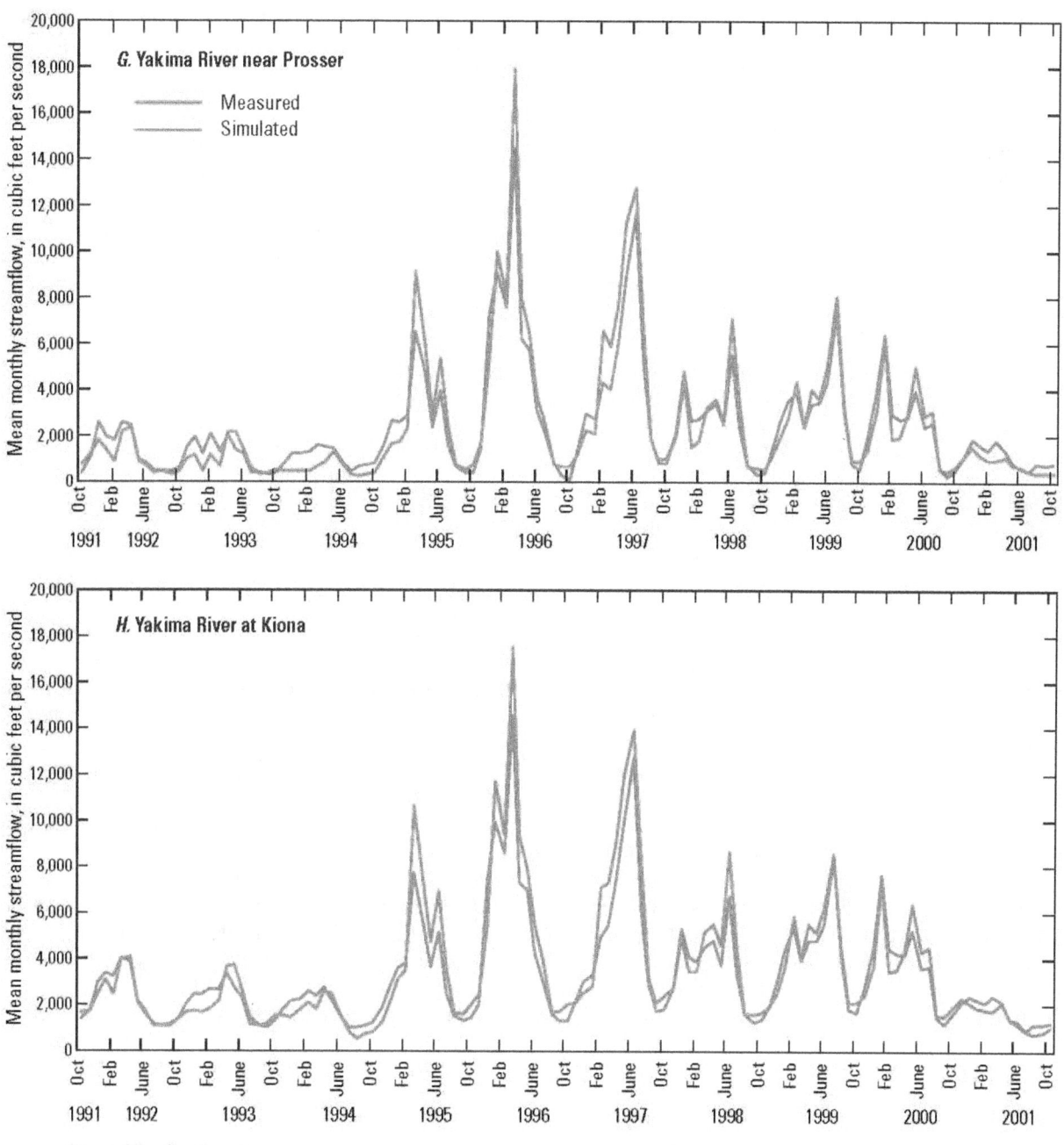

Figure 28.—Continued

By Umtanum, the match is still good but overprediction of low flows is more pronounced because groundwater discharge from the Kittitas basin is occurring throughout the year and not principally in the irrigation season. That is, the modeled streams/drains in the Kittitas basin are not capturing enough of the groundwater discharge (recharge) during the April through October period and tributary flow is underpredicted. This partly is due to direct surface-water discharge to drains that is not simulated by the ECM, and is accounted for in the ECM by direct groundwater discharge. Thirty-three miles downstream at Union Gap (an additional 1,885 mi^2 of drainage area), the simulated and measured hydrographs are similar. The good match is consistent with the ECM's ability to simulate the Naches River at its mouth with a reasonable degree of accuracy (fig. 28D). However, the overprediction of low flows during the winter in the upper basin is propagated downstream, although not as large an extent as at Umtanum. This overprediction is more pronounced in drier years such as 1992–94 and 2001. Some of the factors described above translate into the downstream propagation of uncertainties and thus, potential errors in the ECM-simulated values.

At Parker, the shape of the hydrographs generally is similar and the simulated flows account for two of the major diversions in the basin, one for the Wapato Irrigation Project and the other for Sunnyside Irrigation District. As described above, however, overprediction of winter low flows in dry years occurs, as does underprediction during some months of the irrigation season, especially August. In many cases, simulated flows would need to be within 10 percent at Union Gap in order to meet the two large downstream diversions. The model fit is good 48 mi downstream at Mabton (mean annual simulated streamflow was 91 percent of measured streamflow) and indicates that a reasonable amount of the return flows, which support the increase in streamflow downstream of Parker (especially during the irrigation season), are being captured by the model. Streamflow at Prosser (the next downstream site) is shown because the return flows need to support the major Chandler-Prosser power diversion just upstream of Prosser. Overall, the simulated and measured hydrographs display good similarity at Prosser with some overprediction of winter flows. The irrigation season flows are captured better at Prosser than at Parker, indicating that flows are sufficient to meet the power diversion. By Kiona (16.2 mi downstream of Prosser), there is little bias in the simulated streamflow (over and underpredicted monthly values are about equally distributed), indicating the ECM can be used to assess potential effects of pumping and major management strategies, such as alterations in reservoir releases, on streamflow over some 190 river miles (including the Naches River segment). These effects can be assessed as far downstream as Kiona,

which is the most downstream gaging station in the basin, and effects at the mouth should be reasonably represented because diversions and pumping downstream of Kiona are included in the model. The effects are best analyzed with respect to differences in streamflow between simulations and not actual discharge quantities at selected sites.

Uncertainty in estimated diversion quantities, recharge estimates, and errors in simulated tributary/drain flows generally do not have a large effect on the simulated streamflow. However, errors in simulated tributary/drain flows affect the underprediction of peaks and the overprediction of winter low flows, especially upstream of Parker. The ECM's ability to simulate the tributary/drain flows varied widely (table 7). Simulated mean annual streamflow ranged from 300 percent of measured streamflow at Swauk Creek to essentially no simulated streamflow at Snipes Creek. Wilson and Cherry Creeks are a measure of almost all tributary flow in the Kittitas basin and the model was able to reproduce the quantity and timing of streamflow. A similarity in the simulated and measured variations in the shape and magnitude of the hydrographs over about 190 mi further suggests that the ECM captures the major regional components of the hydrologic system.

The simulated flows for tributary streams and agricultural drains are directly influenced by the simulated heads and the amount of the tributary lower-order branches and drain system included in the SFR2 network. Although a reasonable accounting of the drain and lower-order tributary networks was included in the model (fig. 12), the regional nature of the ECM precluded complete accounting of the networks. A tributary stream may cross two to five mapped HGUs that vary from bedrock units to alluvial units. The drain networks occur in the basin-fill units but they also may cross up to three mapped HGUs. Capturing the general hydrographs for these systems, therefore, would require that the simulated heads, recharge, conductances, and hydraulic properties of the model HGUs all be within reasonable bounds. Heads that are underpredicted greatly affect the resultant streamflow because small head differences in the uplands can be important. The simulated mean monthly values for selected streams/drains generally show that the ECM does not capture the streamflow in some of the upland tributaries (table 7). During model calibration and testing, the model was calibrated only to streamflow observations for selected upland tributaries (each one as a separate calibration run) to determine if heads and thus, streamflow matches could be improved in these areas. Streambed elevations, based on an assumed depth of 2 ft also were adjusted. None of the calibration runs or model testing produced improved tributary streamflow matches. There also are distinct spatial differences in the reliability of the simulated streamflow estimates.

Table 7. Simulated mean monthly streamflow as a percentage of measured or estimated mean monthly streamflow of selected tributary streams and agricultural drains, Yakima River basin aquifer system, Washington.

Month	Swauk Creek	Taneum Creek	Manastash Creek	Wilson-Cherry Creek	Cowiche Creek	Ahtanum Creek	Toppenish Creek	Satus Creek	Sulphur Creek	Spring-Snipes Creek
October	645.8	575.4	109.5	110.9	221.1	134.4	359.0	11.5	137.3	0.0
November	612.1	479.3	132.9	155.7	189.0	169.5	302.8	12.0	211.0	.0
December	596.8	488.8	137.3	143.8	118.6	135.5	195.9	7.4	184.3	2.7
January	409.1	414.9	145.9	132.1	97.4	115.3	116.6	3.0	138.0	.0
February	198.6	265.8	100.5	103.3	90.1	100.1	85.6	1.6	125.8	3.5
March	72.1	111.7	64.2	114.1	91.9	104.8	80.9	1.6	123.7	.5
April	37.4	50.4	35.1	117.8	96.5	97.4	82.4	1.9	109.1	.2
May	47.4	39.7	25.0	78.1	109.5	110.1	117.2	2.7	112.0	.0
June	87.2	63.6	32.0	68.6	118.5	100.1	165.9	4.1	115.7	.0
July	162.0	120.6	45.0	103.6	137.2	170.2	247.3	5.4	125.7	.0
August	291.3	224.1	62.2	70.8	166.7	210.3	293.3	6.3	118.3	.0
September	469.9	392.4	83.4	77.3	188.2	120.9	342.7	7.6	117.5	.0
Average	302.5	268.9	81.1	106.3	135.4	130.7	199.1	5.4	134.9	0.6

Model Uncertainty and Limitations

The ECM represents an extremely complex natural system with a set of mathematical equations that describe the system; this natural system has been perturbed by human activities. Intrinsic to the model is the error and uncertainty associated with the approximations, assumptions, and simplifications that must be made. In addition to those intrinsic errors, hydrologic modeling errors typically are the consequence of a combination of errors in the (1) input data, (2) representation of the physical processes by the algorithms of the model, and (3) parameter estimation during the calibration procedure (Troutman, 1985). These three types of model errors within the ECM and how those errors limit application of the model are as follows:

1. Data on types and thicknesses of mapped HGUs, water levels, pumpage, recharge, and hydraulic properties were taken from Jones and others (2006), Vaccaro and Sumioka (2006), Vaccaro and Olsen (2007a), Jones and Vaccaro (2008), Vaccaro and others (2009), and other sources, and are estimates of actual values. Most of the data were concentrated along the Yakima River valley and populated areas because that is where most wells are located (fig. 4). This means that for some of the study area, information is unavailable to constrain the model, especially for the areas lacking water-level data described previously.

Portions of the model domain include basin-fill sediments, which are unmapped or poorly characterized. In areas without lithologic well logs, variability in hydrogeologic properties or depths of contacts may fall outside the range of values in areas that have been better characterized and the errors associated with this variability would remain unrepresented. Specific conclusions drawn from regions of the model with sparse observations should be limited to general flow directions and relative magnitudes.

Recharge was estimated using physical-process models that preserve a water balance, but uncertainty in input data and parameters of these models, limiting assumptions, and model accuracy can produce error in these estimates. Recharge is the major source of water to the flow system and any errors in recharge generally would affect simulations results throughout the system. The initial hydraulic-property data generally came from specific-capacity tests, which typically measure drawdown at one time and at one pumping rate, and are not as accurate as aquifer tests. Thus, broad ranges of hydraulic-property parameter values are possible, especially for the deeper part of the flow system. The large number and consistency of hydraulic-property parameter values from previous modeling studies helped to constrain the initial estimates. However, uniform K_h values for basin-fill, interbed, and bedrock model HGUs resulted in simplifications to spatially varying properties. Lack of information on streambed hydraulic conductivity values resulted in these values being poorly constrained and may limit the accuracy of groundwater/surface-water exchanges.

The uncertainties in the prescribed flows for smaller creeks and diversions can affect the accuracy of the ECM in some areas. Measured streamflow for most of the tributary streams and agricultural drains was not available for the simulation period, and therefore, the model could not be constrained by measured data at these locations. However, the use of monthly mean flows for selected streams calculated from the watershed models helped to constrain parameters, although there is uncertainty in these modeled values. There are more than 2,000 mi of canals, laterals, and drains in the basin, and only the major parts of this complex system were simulated. Locally, the exclusion of smaller drains limits the testing of management alternatives relative to analysis of irrigation-return flows. The ECM simulation period ends in September 2001, and this may be a limitation for some uses of the model.

In areas with sparse or no observation data, the parameters are less constrained and have greater uncertainty than parameters in areas with adequate observation data. In these areas, the potential uses of the ECM would be limited. This limitations increase with depth because the number of head observations decrease with depth, especially beyond layer 12 (fig. 20). Where observations are available in deeper layers, they tend to occur in the lower basin, and thus, the model would be limited for assessing management alternatives deeper in the system in the upper basin, for example in the northern part of the model domain. The lack of observations in the bedrock units, unknown vertical variability in the hydrogeology of these units, and the relative insensitivity of the bedrock hydraulic properties indicate that ECM applications should be limited to the non-bedrock area of the model domain.

2. A numerical model cannot completely represent all physical processes within a watershed. Determining if a weakness in a simulation is attributable simply to input data error or shortcomings in how the model represents the governing physical processes is intractable. The model inevitably relies on simplifying assumptions and generalizations that complexly affect the results of the simulation. The ECM was not designed to represent every detail of the hydrologic system (a task beyond the scope of the investigation), and simulation results will vary based on which details were and were not included. For example, small differences in simulated heads in the upland areas can result in large differences in simulated tributary streamflow because the complex nature of the stream system and valleys in these areas was not represented in the model. Model-discretization errors result from (1) the effects of averaging elevation information over the model cell size, (2) the time-averaging of observed values inherent in a monthly simulation stress period and (3) the inaccuracies in the geometric representation of mapped HGUs (especially in the representation of the bedrock areas and its contact with the basalt and basin-fill units). For this reason, interpretations of simulation results should be limited to scales several times greater than the model spatial and temporal resolutions of 1,000 ft and 1 month, respectively or larger in areas of steep surface gradients or rapidly changing conditions.

3. Errors in parameter estimates occur when improper values are selected during the calibration process. Various combinations of parameter values can result in low residual error, yet improperly represent the actual system. An acceptable degree of agreement between simulated and measured values does not guarantee that the estimated model parameter values uniquely and reasonably represent the actual parameter values. The use of automatic parameter estimation techniques and associated statistics, such as composite scaled sensitivities and correlation coefficients, removes some of the effects of non-uniqueness, but certainly does not eliminate the problem entirely. The comparison of calibrated values to literature values also can reduce error caused by parameter estimation if the model results are within previously accepted ranges. Limitations of the observation weighting scheme used in this study include non-varying weights for heads that did not take into account errors in simulated heads for multilayer wells, uncertainty in the model HGU designations, and head errors in possible perched water-table areas, and streamflow weights not based on sample or measurement error (streamflow variance).

If the regional groundwater-flow model is used appropriately, the effects of the simplifications and other potential errors can be limited. If the model is used for simulations beyond which it was designed, however, the generalizations and assumptions used could significantly affect the results.

Appropriate Uses of the Groundwater Model

Because the ECM was constructed to simulate regional-scale groundwater flow, it can be used to help answer questions regarding groundwater-flow issues at that scale. For example, interactions can be considered between hydraulic heads, discharge, pumping, and flow direction and magnitude on a regional scale. The ECM model, however, also includes many local-scale features and site-specific data. Examples of such features include small grid cells, numerous model HGUs (including 20 defined for the basin-fill sediments), flow barriers, diversions and returns, streamflow routing, and pumpage from many wells.

The ECM can be used to evaluate alternative conceptualizations of the flow system that are likely to have a regional effect. These might include the effects of changes in recharge caused by climate change, different interpretations of the extent of geologic structure, changes in reservoir operations, or other conceptual models that would affect the spatial variation of hydraulic properties (Faunt and others, 2004).

The model also can be used to provide insight about travel/residence times and thus, provide an improved understanding of the potential pathways for contaminant transport. Flow direction and magnitude may be appropriately represented using particle-tracking methods as long as the particle paths are interpreted to represent advective-transport flow paths that are at least several times longer than the length of a 1,000-ft model cell (Tiedeman and others, 2003). That is, flow paths and travel/residence times can be assessed over the lengths of the structural basins. Such information could provide generalized information for such aspects as the pathways and residence times for nitrate entering the groundwater system at different locations.

The monthly stress period used in the model was selected to provide adequate temporal resolution for analyses of variations in pumping and recharge rates corresponding to seasonal changes in precipitation and irrigation. Impacts of management decisions on time scales less than 1 month are unlikely to be adequately simulated with the ECM, but the model can be used appropriately to simulate seasonal or annual changes to water-use practices, crop types, or potential future climate.

The basic structure of the calibrated ECM allows for alternative uses of the model. Sensitivities of stresses on the flow system can provide information for directing additional data analyses and (or) data collection. Cause and effect also can be assessed. For example, assessing the effect of a 20 percent decrease in recharge or prescribed streamflow on different processes in the flow system would be an appropriate use of the ECM. For all applications, simulated results should be compared to heads or flows simulated using the calibrated ECM and not actual simulated heads (elevations or depth to water) or flows. For example, it might be said that groundwater-level declines in one location may be on the order of 50 ft due to increased pumping, but it would not be appropriate to state the projected water-level altitudes in wells at the same location. The model is best suited to analyze potential changes in various aspects of the flow system relative to the calibrated ECM.

The model can be used for examining the effects of continued or increased pumping on the regional groundwater-flow system to effectively manage groundwater resources. For increased pumping, the model should not be used for assessing effects of one or two new wells on the flow system but could be used for such applications as estimating the quantity of pumpage in an area that leads to unacceptable groundwater-level declines and (or) streamflow capture. An example application of continued ECM pumping projected into the future is presented below. With increasing demand for water in the basin, the ECM could be further developed to test optimization strategies for the conjunctive use of surface water and groundwater given specified constraints such as minimally allowable streamflows during certain times of the year (such as August) at selected points along the Yakima River.

Model-Derived Groundwater Budget

The ECM can be used to derive components of the groundwater budget for each of the 504 stress periods for the 42-year simulation. During this period, the distribution and amount of pumpage changed and recharge varied, and a cumulative or mean annual water budget would not highlight these variations. Thus, simulated water budgets are presented for a wet (1997), average (2000), and dry (2001) year. These three years capture the hydrologic variability present in the model domain and are representative of the existing conditions. As measured by the Yakima River at Kiona (the most downstream streamflow site in the basin), the ratios of the annual mean discharge (sum of the daily mean discharges for one year divided by the number of days in that year) to the mean annual discharge (sum of the annual mean discharges divided by the number of years; calibration period of 1960–2001) for 1997, 2000, and 2001 were 1.74, 1.08, and 0.44, respectively. These ratios show that these years are representative of a wide range in climatic conditions, and thus also a wide range in recharge, reservoir operations, surface-water delivery, and pumping.

The simulated water budgets (fig. 29) show the variations in inflows and outflows between the three types of climatic years. Outflows are fluxes out of the aquifer system. Inflows are fluxes into the aquifer system. For example, recharge is considered an inflow and pumping wells are considered an outflow. Groundwater recharge, which drives the system, ranged from about 3,200 ft³/s in 2001 to 9,700 ft³/s in 1997, a 200 percent increase compared to a dry year. In 2000, recharge was within 3 percent of the mean annual recharge for the calibration period. Recharge to the system is reasonably matched by the outflows from the aquifer system into storage for 1997, 2000, and 2001. In 2001, less water flowed out of the groundwater-flow system into storage because water levels were lower in the dry year (2001) than in the wet (1997) and average (2000) years. The next largest changes between years are inflows to the flow system from storage. In 1997 and 2000, inflows from storage are nearly the same, but inflows from storage in 2001 are about 600 ft³/s larger than in 1997 and 2000. For the dry year (2001) with less recharge and additional effects of pumping (wells) than in the wet and average years, the inflow of water into the groundwater-flow system from storage is clearly shown in the outflows minus the inflows budget, where the 2001 difference is an inflow of about 2,900 ft³/s.

The next largest budget component is the drain flow (fig. 29), which is the boundary condition that accounts for the groundwater discharge supporting streamflow in the humid uplands. Groundwater discharge to drains ranged from nearly 3,500 ft³/s in 2001 to nearly 4,000 ft³/s in 1997. The variations by type of climatic year are well represented and captured by the ECM for this water-budget component. Groundwater discharge to streams (stream leakage outflows) is the next largest component after drain outflows. These values ranged from about 2,200 ft³/s in 2001 to 2,600 ft³/s in 1997, and they are about six times larger than the streamflow losses (stream leakage inflows). The head-dependent boundary flows show that there are both flows into and out of the system with a net inflow that ranges from about 202 ft³/s in 1997 (higher groundwater levels in the wet year result in smaller inflows) to 260 ft³/s in 2001 (lower groundwater levels in the dry year result in larger flows); these flows are sensitive to calculated heads in the uplands. Wells are the next largest budget component at about 400 ft³/s, ranging from about 395 ft³/s in 1997 and 2000 to 460 ft³/s in 2001. The constant-head boundary along the Columbia River receives groundwater discharge at about a constant rate (255 ft³/s) during the three years, and any flow provided to the system was simulated to be relatively small and nearly constant at about 30 ft³/s. Many of the differences between the three years are consistent with interannual differences in water levels.

The net annual budgets (outflow minus inflow) (fig. 29) for each of the three representative years shows that recharge is the largest water-budget component and dominates the inflows to the system. Recharge primarily is balanced by outflows from the groundwater system into storage. A part of recharge is balanced by discharge (drains) in the humid uplands to perennial streams and groundwater discharge (stream leakage) to the major tributary streams and rivers. The head-dependent boundary flows (reservoirs) provide a net inflow to the system of about 230 ft³/s and the Columbia River provides a net outflow from the system of about 225 ft³/s.

Model Applications

The ECM was used to estimate the response of the regional flow system to potential changes in stresses. These scenarios are used to better understand the relation of the groundwater-flow system to surface-water resources. In particular, the scenarios simulate the relation between pumping and surface-water resources. The potential effects of scenarios are assessed by comparing simulated output from the scenarios with simulated output from the calibrated ECM; that is, streamflow simulated from a scenario is compared to the ECM streamflow along the SFR2 network, which does not include streamflow in the drain network. All assumptions and limitations underlying the ECM are valid for the scenarios.

The long (42-year) simulation period for the ECM (October 1959 to September 2001, water years 1960–2001) allows for a temporal assessment that accounts for changes in pumping over time. Explicitly included in the ECM is a large range in climatic conditions and thus, unregulated runoff and natural recharge. Reservoir operations and diversions in the basin also have changed throughout this period, leading to different quantities of surface water entering and leaving the modeled river system. As a result, a broad range of hydrologic conditions (both natural and human induced) are simulated in the ECM, which in turn, is represented in the simulated streamflow for about 200 mi of the Yakima River, 17 mi of the Naches River, and numerous tributaries and agricultural drains.

Five model scenarios were simulated using the ECM. Each scenario included existing-condition recharge of the ECM, except scenario 3, which included additional estimated recharge from septic systems. The first three scenarios use the same 42-year period and initial hydraulic heads for all model layers as the ECM. The scenarios are:

1. **Existing conditions without pumping.**

   This scenario was formulated to better understand how the relation between historical pumping and surface-water resources has changed over time. All groundwater pumpage was eliminated for this scenario.

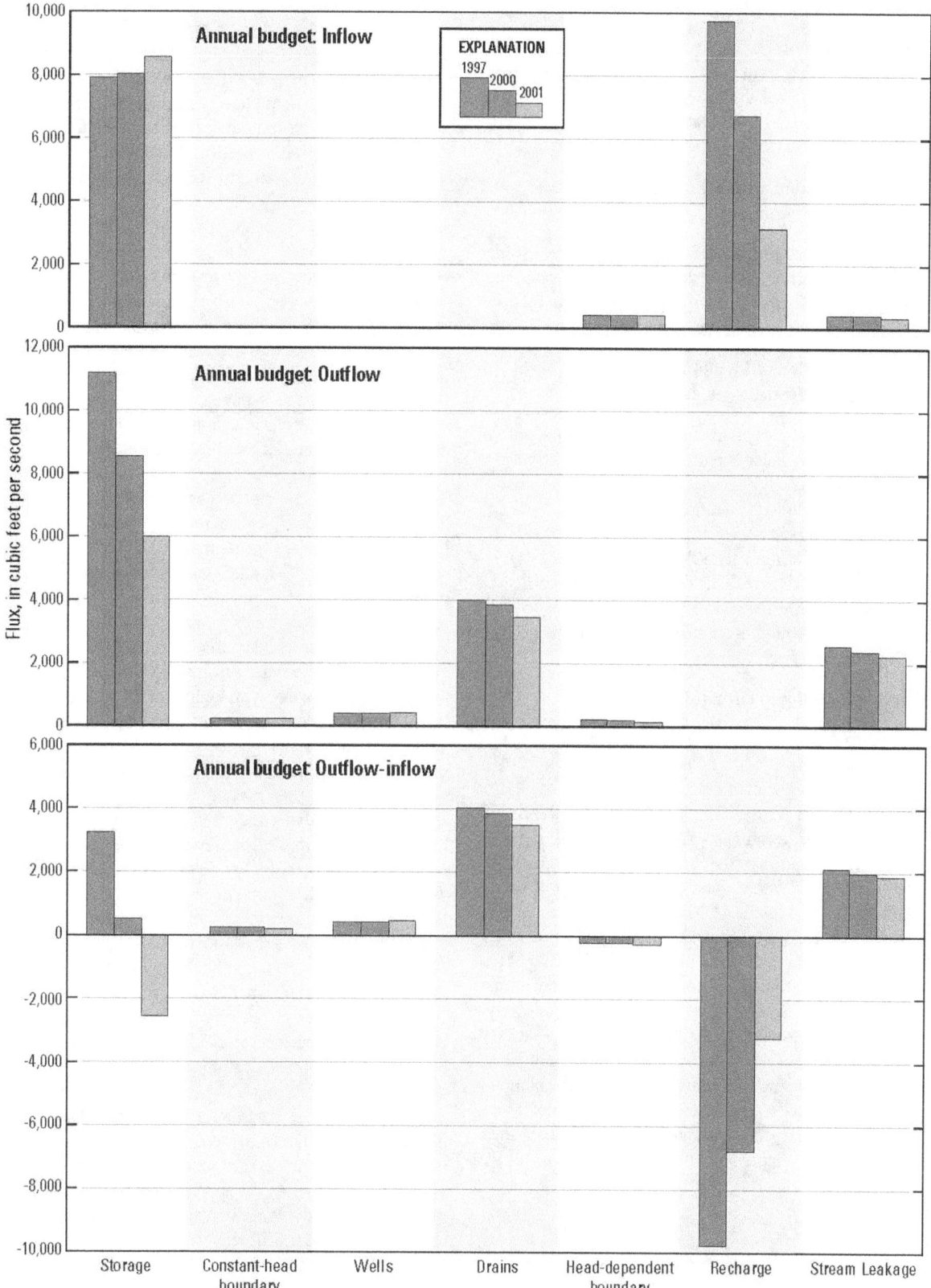

Figure 29.   Simulated annual water budgets for a wet (1997), average (2000), and dry (2001) year, Yakima River basin aquifer system, Washington.

2. **Existing conditions without basalt pumping.**

The purpose of this scenario was to assess the simulated effects of pumping from the basalt units on streamflow. All pumping from basalt was eliminated for this scenario. Pumping for selected wells that withdraw water from both basin-fill units and basalt units were not eliminated.

3. **Existing conditions with additional recharge and without exempt pumping.**

This scenario simulates the effects of pumping by exempt wells on streamflow. All estimated exempt-well pumping was eliminated for this scenario and compared to the ECM with additional recharge from septic-system returns.

4. **Existing-conditions with additional pumping from pending groundwater right applications.**

This scenario includes additional pumping estimated for 886 pending groundwater right applications in the basin. (Five groundwater right applications for Frost Protection use were not included.) Additional groundwater-application pumping was simulated for October 1993–September 2001. Simulated values from the ECM are compared to those calculated from this scenario. The scenario addresses what the potential impact on streamflow and water levels may have been if the pending applications had been approved.

5. **Existing conditions continued into the future.**

This scenario simulates potential future conditions using existing boundary conditions, including pumping for October 1993–September 2001 (water years 1994–2001). These last 8 years of the ECM simulation period were repeated 3 times, creating a single simulation of 66 years (42-year period of the ECM plus the additional 24 years). Simulated future streamflow was compared to the last 8 years of the ECM. This 8-year period includes dry, average, and wet years and thus, a large range in reservoir operations, including accounting for Title XII operations. Twenty-five percent of this period consisted of prorating years (1994 and 2001). Thus, this period likely is representative of near-term future conditions because 26 percent of the water years from 1977 through 2010 were prorating years. Excluding prorating years, pumping also was estimated to have increased by only about 2 percent during this period, and thus, is nearly constant for non-prorating years.

The scenarios meet the purpose of the study: to provide an improved understanding of the groundwater-flow system and its relation to surface-water resources during all or part of the 1960–2001 period. Differences in scenario streamflow relative to the ECM-simulated values provide the improved understanding of the source of water supplied to wells in the basin. These simulated effects also provide information for assessing the groundwater resources because potential future trends in groundwater levels, and thus streamflow, are important for understanding groundwater availability. Trends in water levels, and absence of trends, when analyzed in conjunction with groundwater pumping and streamflow, can indicate areas of (1) changes in groundwater storage, (2) potential capture of recharge, and (3) changes in groundwater discharge to streams, springs, or wetlands. Implicitly included in (2) is the potential to increase streamflow losses due to increases in pumping. Basic laws of physics require that changes in pumping must be balanced by one or a combination of the three changes listed (Theis, 1940).

The direct comparison of simulated scenario values to ECM values provides that understanding in a limited sense because there is multiple variations in data among scenarios. For example, recharge from groundwater irrigation was the same for scenarios 1, 2, and 5, additional recharge from return septic flows was included in scenario 3 and additional recharge from groundwater irrigation was not included in scenario 4. The direct comparison of simulated scenario values to ECM values are sensitive to simulated groundwater levels because in areas with higher than measured water levels, the simulated groundwater discharge to streams may be larger than what actually occurs. As a result, diminished groundwater discharge due to pumping may have a larger net effect on simulated streamflow. Conversely, in areas with simulated water levels lower than measured water levels, there may be no simulated discharge to the SFR2 network and simulated streamflow loss in these areas would be less than what may occur. Similarly, the streamflow differences would be sensitive to the simulated streamflow stage. For example, if a groundwater-level pumping response of 1 ft occurs in February in an area with simulated streamflow losses, the losses would be much smaller at a discharge quantity of 2,000 ft^3/s than at a discharge of 15,000 ft^3/s. The reason for smaller losses is because streamflow gains and losses are linearly related to the elevation difference between the water table and the stream stage. The estimated pumping has an error band (Vaccaro and Sumioka, 2006) and thus, any errors in pumping would translate to a relatively proportional error in simulated differences in streamflow. In addition, estimated pumping was less than the appropriated quantity, indicating that if the appropriated rights were used for the pumping stress, the simulated differences in streamflow from the ECM results would be much larger.

The simulated scenario results do not directly address streamflow depletion, which generally is considered the quantity of water that a pumping well obtains from a stream. Although a well may obtain all of its water from a stream (100 percent capture), if some or all that water is returned to either the groundwater or surface-water system, the scenario streamflow loss would be smaller than the actual localized streamflow depletion.

Unless otherwise stated, the streamflow comparisons for the scenarios are for Yakima River at Horlick (Horlick), Yakima River at Umtanum (Umtanum), Yakima River above Ahtanum Creek, at Union Gap (Union Gap), Yakima River near Parker (Parker), Yakima River at Mabton (Mabton), Yakima River at Prosser (Prosser), Yakima River at Kiona (Kiona), Yakima River at Richland (Richland), and Naches River at North Yakima (Naches) (table 5). Except for Richland, all sites are located at existing gaging stations. Other comparisons are described for two other locations defined by river miles: river mile (RM) 98 (upstream of Toppenish) and RM 83 (near Granger). Scenario results for Horlick are less reliable because some of the effects on streamflow are due to pumping in bedrock units, and these units have not been mapped in detail. Scenario results are not shown for Yakima River at Cle Elum for the same reason. Several sites capture the total upstream effects of pumping relative to a structural basin: Horlick for the Roslyn basin, Umtanum for the Kittitas basin, Union Gap for the Selah/Yakima basins, Prosser for the Toppenish basin, Kiona for the extended Toppenish basin, Richland for the Benton basin, and Naches for part of the Naches River and Selah basins. The comparisons are from 1960 through water year 2001. Although pumping has not increased much relative to the total pumping since 2001, increased pumping after 2001 may be locally important. The terms difference and differences (scenario values minus ECM values) are used in the following sections in the comparison of the scenario results to the ECM results.

Scenario 1—Existing Conditions without Pumping

Streamflow for the 42-year period was simulated using the ECM operated without groundwater pumping. Total pumping during this period that was eliminated was estimated at about 9.7 million acre-ft (mean annual pumpage of 320 ft^3/s or 231,200 acre-ft), and ranged from a minimum of about 114,000 acre-ft (160 ft^3/s) in 1960 to 370,000 acre-ft (510 ft^3/s) in 2001.

The difference in simulated streamflow over the 42-year period increases in a downstream direction (fig. 30). For example, the annual difference at Umtanum was 18 ft^3/s in 2001, and it increased to 194 ft^3/s at Richland in 2001. The annual difference did not increase linearly in a downstream direction because of system complexity due to mainstem streamflow quantity, tributary inflows, diversions, and returns. For example, in 2001 the difference between calibrated ECM and scenario 1 streamflows was about 75 ft^3/s at Union Gap (RM 107.3), it increased to about 126 ft^3/s at RM 98, and by Mabton (RM 59.8), it was about 168 ft^3/s. The simulated differences increase over time, but display less variability (especially for sites upstream of Parker) after about 1995 due to new groundwater withdrawals primarily occurring for domestic use and not for uses with a high water demand, such as irrigation. For all sites, annual differences ranged from less than 1 ft^3/s at Horlick in 1960 to 224 ft^3/s at Richland in 1995; the differences at Umtanum and Naches were similar throughout the simulation period, and ranged from about 2 ft^3/s in 1960 to 23 ft^3/s in 1995. At Richland, the annual difference represents about 38 percent of the total pumpage in 2001 (a prorating year). In comparison, in 2000 (an average non-prorating year), the annual difference at Richland was about 48 percent of the total pumpage. The percentage was larger in 2000 because the pumpage was less and the effects of basalt pumpage propagate through the system at different temporal resolutions.

The mean annual, in comparison to annual mean, difference in simulated streamflow at Kiona between the calibrated ECM and scenario 1 was 143 ft^3/s (table 8). This was an increase of about 4 percent or equivalently about 46 percent of the mean annual pumpage in the ECM model. Further indicating that about 54 percent of the pumpage in the ECM model was principally met by water released from aquifer storage. Some pumpage demand also would have been met by other factors, such as increased recharge or decreased spring flow. Relative to the total cumulative water budget for the 42-year period, eliminating total pumping changed the total inflows to model cells by about 1 percent through aquifer storage changes and changed outflows by about 1 percent through increases in discharge to streams.

On a monthly basis, the largest effects generally occur from July to August for sites upstream of RM 98. At RM 98, the largest effects (seasonality) were more spread out over time. Thus, the effect of pumping becomes more variable starting at about RM 98 (principally due to basalt pumping). Overall, monthly differences in streamflow exceeded the annual differences. The largest monthly difference in streamflow (265 ft^3/s) was at Richland in 1995, and the largest difference in 2001 was about 235 ft^3/s at Richland.

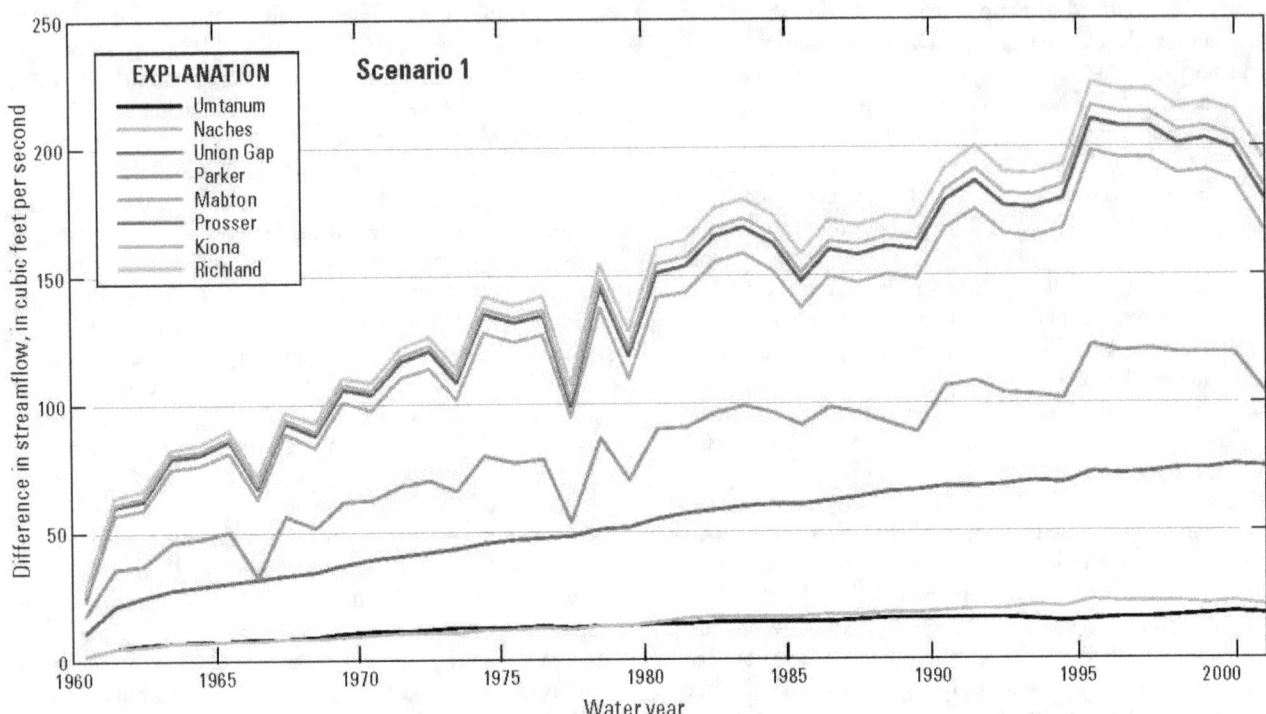

Figure 30. Difference in simulated mean annual streamflow between existing conditions and existing conditions without groundwater pumping, Yakima River basin aquifer system, Washington.

Table 8. Simulated mean annual groundwater pumping and streamflow for base case, and mean annual change in groundwater pumping and streamflow from base case for model scenarios, 1960–2001, Yakima River basin aquifer system, Washington.

[Negative values represent a decrease from base case. Positive numbers represent an increase from base case. **Abbreviation**: ft³/s, cubic foot per second]

Scenario No.	Description	Groundwater pumping (ft³/s)	Change in pumping (ft³/s)	Streamflow (ft³/s)		
				Yakima River at Horlick	Yakima River at Umtanum	Yakima River above Ahtanum Creek at Union Gap
Base case[1]	Calibrated model	320		2,187	2,534	3,605
Scenario 1	No pumping	0	-320	0	13	53
Scenario 2	No basalt pumping	185	-135	0	1	8
Scenario 3	No exempt pumping	290	-30	0	1	2
Scenario 4[2]	Groundwater right applications	650	330	1	-10	-23

Scenario No.	Description	Streamflow (ft³/s)			
		Yakima River near Parker	Yakima River at Prosser	Yakima River at Kiona	Naches River at North Yakima
Base case[1]	Calibrated model	2,283	2,409	3,408	1,426
Scenario 1	No pumping	82	139	143	15
Scenario 2	No basalt pumping	15	25	26	4
Scenario 3	No exempt pumping	3	7	7	1
Scenario 4[2]	Groundwater right applications	-31	-58	-62	-6

[1] Base case values represent simulated rates in calibrated model.

[2] Scenario 4 run for water years 1994–2001 only.

Scenario 2—Existing Conditions without Basalt Pumping

For this scenario, the ECM was simulated for the 42-year period without the estimated basalt pumping. The mean annual pumpage for this scenario was 133,200 acre-ft (or 185 ft³/s, table 8); a 42-percent decrease from ECM's 231,200 acre-ft. Decreases in pumpage ranged from about 25,000 acre-ft (22 percent of ECM pumpage) in 1960 to 188,000 acre-ft (51 percent of ECM pumpage) in 2001. Most of the measured groundwater-level declines in the basin have occurred in the basalt units (Vaccaro and others, 2009), suggesting that basalt pumpage may principally be met by changes in aquifer storage. This implies that if basalt pumpage is decreased in the future, the effects of current withdrawals will continue into the future.

The annual differences in simulated streamflow between the ECM and scenario 2 (fig. 31) at Richland ranged from about 3 ft³/s in the early years to about 57 ft³/s in 1995. There is a distinct downstream increase in the difference in simulated streamflow because most of the basalt pumping occurs in the lower basin. For example, the largest difference in simulated annual streamflow at Umtanum was about 3 ft³/s, whereas the largest differences at Union Gap and Parker were about 14 and 29 ft³/s, respectively. By Kiona (77.4 river miles downstream

of Union Gap), the largest difference increased to 53 ft³/s. As a percentage of the ECM streamflow, the mean annual difference at Kiona was 1 percent (26 ft³/s, table 8), and this difference is about 19 percent of the mean annual ECM basalt pumpage.

Relative to total basalt pumpage in 2000 (non-proratable) and 2001 (proratable), about 16–17 percent of the basalt pumping for each year was met by a decrease in groundwater discharge to streams and an increase in streamflow losses— the streamflow difference between scenario 2 and the ECM. As previously discussed, available data indicate that much of the pumping was met by changes in aquifer-storage (groundwater-level declines) during the 42-year period. This is in contrast to the effects of eliminating total pumping in the basin, which is primarily from basin-fill units (scenario 1).

To examine the differences between pumping from the basalt and sediment layers, two additional simulations were examined. The ECM was simulated with no pumping from October 1959 to September 1980, then pumping only from the basalt HGUs (fig. 32A) from October 1980 to September 1994, and then no pumping for the remainder of the simulation period. The process was repeated with pumping only from the sediment HGUs (fig. 32B). The additional simulations show that the pumping from the basin-fill units has a stronger relation to surface-water resources than basalt pumping (fig. 32).

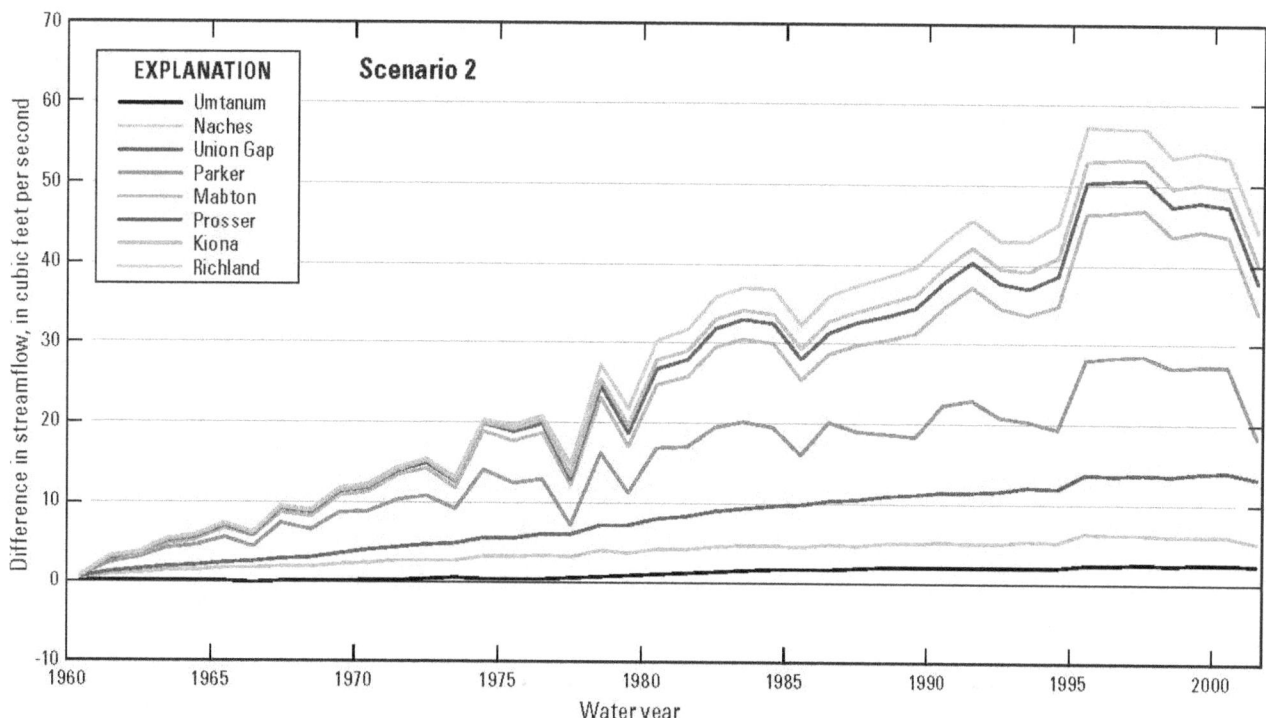

Figure 31. Difference in simulated mean annual streamflow between existing conditions and existing conditions without basalt pumping, Yakima River basin aquifer system, Washington.

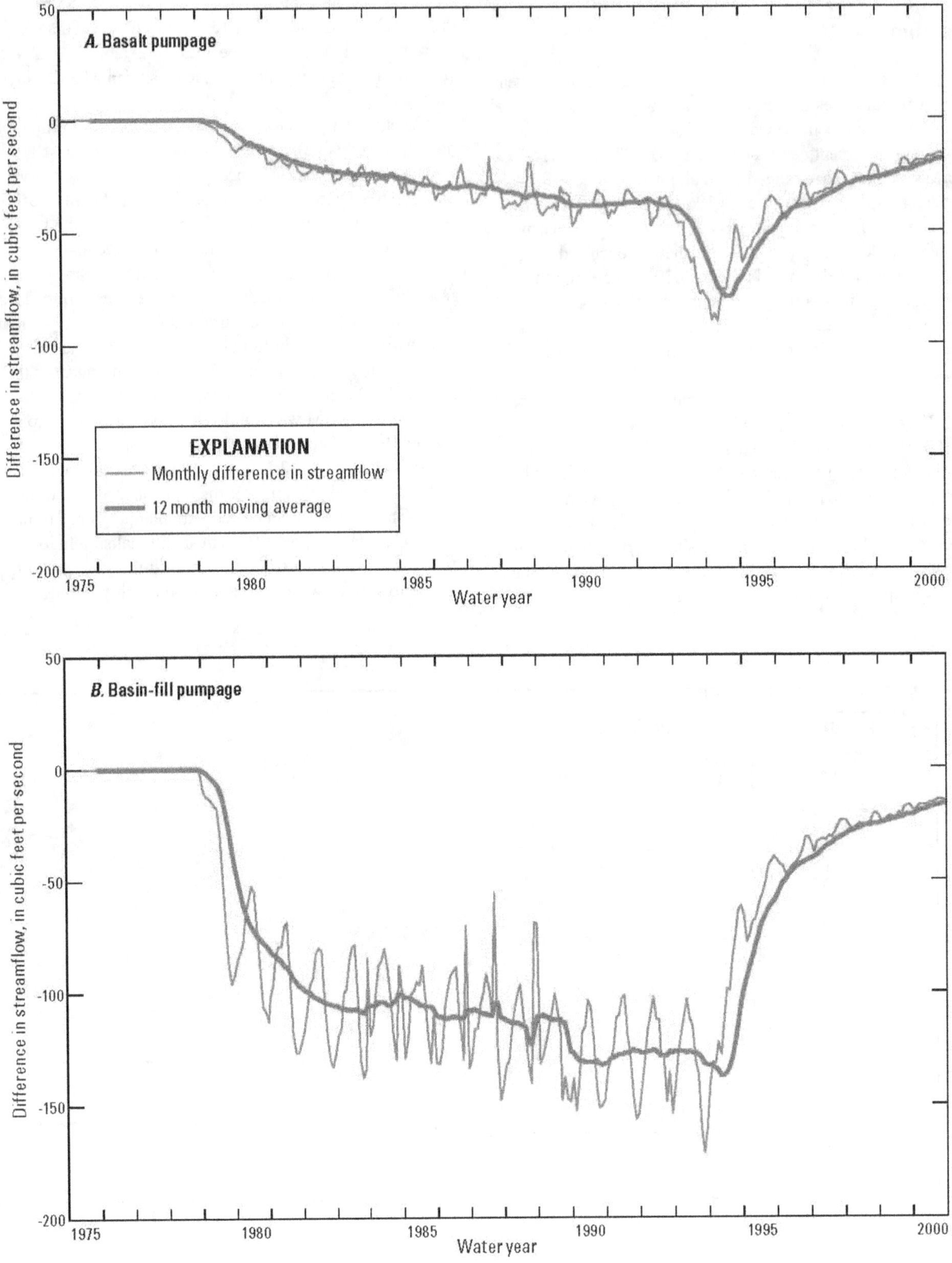

Figure 32.   Difference in simulated mean annual streamflow between existing conditions and existing conditions with only basalt pumpage from 1980 to 1994, and existing conditions with only basin-fill pumping from 1980 to 1994, Yakima River basin aquifer system, Washington.

Scenario 3—Existing Conditions with Additional Recharge and without Exempt Pumping

The ECM was used to simulate the 42-year period without the estimated exempt pumping for scenario 3. The mean annual pumpage was 209,400 acre-ft (290 ft³/s), which is a 9-percent decrease (21,800 acre-ft) from the ECM (table 8). Exempt pumpage (self-supplied domestic and most Group B public water supply systems) was about 9 percent of the total pumpage in 2000 (Vaccaro and Sumioka, 2006). Domestic exempt pumpage was represented by 3,718 wells centered in census blocks from 2000, and 84 percent of those wells were finished in the basin-fill units, and public water supply exempt pumpage was represented with 1,022 points of withdrawal (83 percent was from the basin-fill units). For this scenario, the ECM was first operated with additional estimated recharge from septic systems and the simulated streamflows were then used as the base case for assessing the relation between exempt pumpage and surface-water resources. Septic return was accounted for in this scenario because the total exempt pumpage was not large relative to the other scenarios and a larger part of this pumpage is returned to the system compared to other use categories. Septic recharge was estimated based on a modified method outlined in Vaccaro and Olsen (2007a). For this study, a base-level percentage of 5.7 percent (in contrast to Vaccaro and Olsen's 5.5 percent) of pumpage for indoor use was used for the months of March through October, resulting in an additional mean annual septic

return value of about 1,515 acre-ft (2 ft³/s). The septic return was input into the ECM as an injection well located in layer 1 at the cell location (row-column) of each exempt well. The septic return was not simulated when the exempt pumpage was not simulated.

Annual differences in streamflow were small (fig. 33). The differences between simulated streamflow in scenario 3 and the calibrated ECM with additional recharge were less than 1 ft³/s at Horlick, and at Umtanum and Naches, the differences were about 1 ft³/s. Annual differences in streamflow at Union Gap and Parker generally follow the same pattern, with differences at Parker being larger by about 2 ft³/s (fig. 33). All values at Parker were less than 6 ft³/s. The annual difference pattern at Mabton, Prosser, Kiona, and Richland are similar. Differences in streamflow at these sites also display more interannual variability than the upstream sites. During the last 10 years of the simulation, annual differences at these sites generally ranged from 8 to 10 ft³/s. Mean annual differences in streamflow generally were not large (table 8), which is consistent with the smaller quantity of exempt pumpage. In 2000, about 26 percent of the pumpage was met by a change in streamflow.

During model calibration and testing, differences in streamflow attributed to exempt pumpage were sensitive to basin-fill parameter adjustments, suggesting that local variations in hydraulic properties may significantly influence the differences in streamflow in ways not represented in this regional-scale model. This sensitivity is due to the fact that most of the exempt pumpage occurs from the basin-fill units.

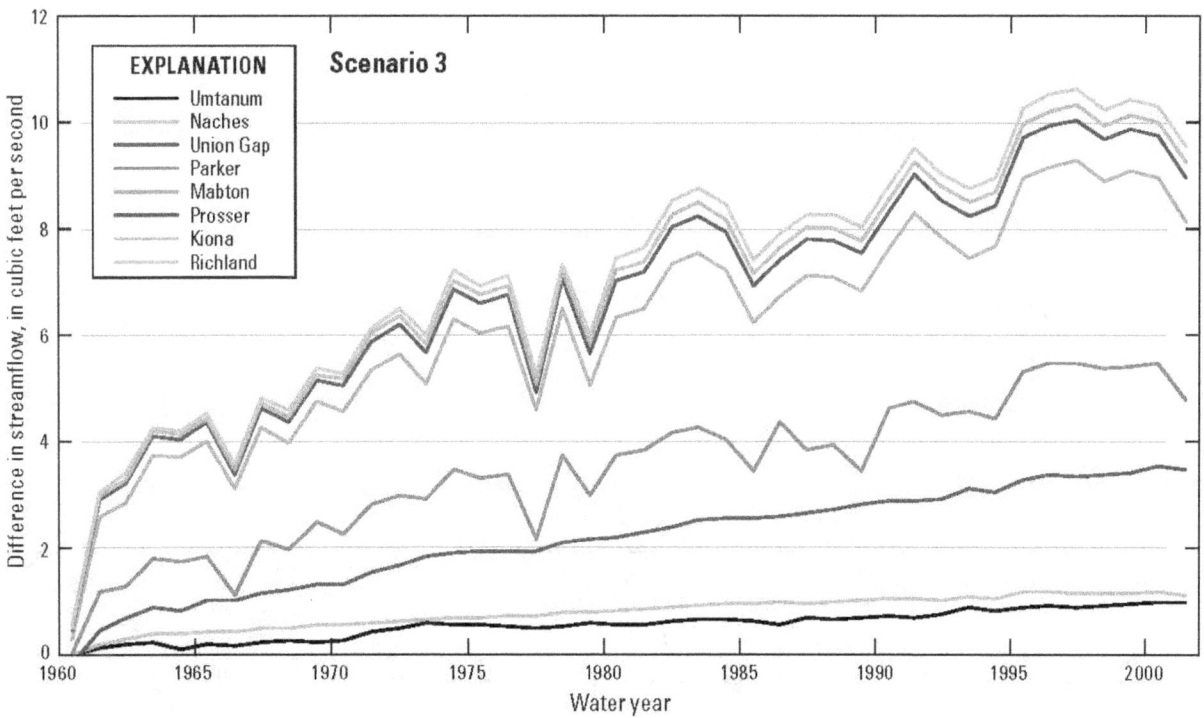

Figure 33. Difference in simulated mean annual streamflow between existing conditions and existing conditions without exempt pumpage, Yakima River basin aquifer system, Washington.

Scenario 4—Existing Conditions with Additional Pumpage Estimated for Pending Groundwater Applications

Estimated pumpage for 886 pending groundwater applications (described in appendix B) is included in this scenario. An additional 237,000 acre-ft (330 ft³/s) of annual pumpage (a quantity referred to as GWAP herein) was estimated for the applications under the assumption that the applications would have been approved for the requested use. For model simulation, GWAP was assumed to start in 1994 and continue through 2001. This 8-year period includes dry, average, and wet years and, as described previously, likely is representative of a range of near-term future conditions. The model hydrogeologic units assigned to the applications were about equally divided between basin-fill and basalt units, but the quantity distribution was much different: 16 percent of the GWAP was from basin-fill units, 15 percent from the Saddle Mountains unit, 40 percent from the Wanapum unit, and 29 percent from the Grande Ronde unit. The GWAP is similar to the 42-year mean annual pumpage for existing conditions (table 8). For this scenario, pumpage ranged from 543,000 acre-ft (750 ft³/s) in 1995 to 607,000 acre-ft (840 ft³/s) in 2001 compared to the ECM values of 306,000 acre-ft (425 ft³/s) in 1995 and 370,000 acre-ft (510 ft³/s) in 2001.

The simulated potential effects of the GWAP on streamflow (fig. 34) compound those estimated in scenario 1 (fig. 30). After 8 years, effects at Umtanum and Naches were decreases of about 14 and 7 ft³/s, respectively. The effects of GWAP are a 37 ft³/s decrease in streamflow at Parker in 2001 and at Prosser, there was a simulated 73 ft³/s decrease (similar to that at Mabton and Kiona). In 2001, this decrease was about 91 ft³/s at Richland (fig. 34). Mean annual decreases in streamflow follow the annual pattern of increasing in a downstream direction (table 8). Annually, between about 6–29 percent of the annual GWAP (330 ft³/s) is met by streamflow decreases. Although GWAP was about 73 percent of the ECM 8-year average pumpage, the percentage of scenario 4 pumpage met by differences in streamflow was not as large as the percentage of scenario 1 pumpage met by differences in streamflow because much of the scenario 4 pumpage occurs from the deeper basalt units that are relatively more isolated from surface-water resources. As a result, scenario 4 had a net loss in aquifer storage because 84 percent of the GWAP pumpage was from the basalt units. As the system approaches equilibrium under the GWAP, aquifer-storage changes would decrease and streamflow would continue to decrease. It is unknown how long it would take for the groundwater-flow system to reach equilibrium under GWAP.

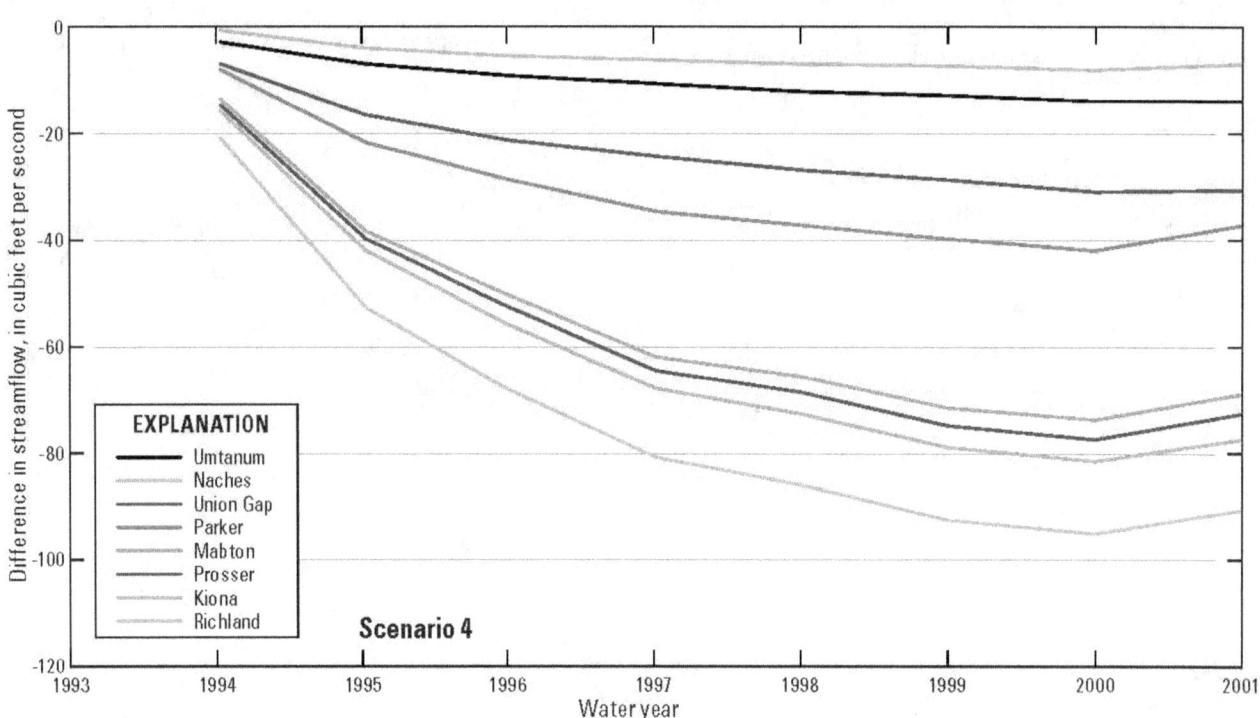

Figure 34.   Difference in simulated mean annual streamflow between existing conditions and existing conditions with additional pending groundwater application pumpage, Yakima River basin aquifer system, Washington, 1994–2001.

Scenario 5—Existing Conditions Continued into the Future

The ECM was used to simulate future possible effects of existing pumpage on the groundwater-flow system and its relation to surface-water resources. The simulation used existing boundary conditions, including pumpage, for October 1993–September 2001. As described previously, this period likely is representative of a range of near-term, future conditions and preserves the use of standby/reserve groundwater rights. During this 8-year period, mean annual pumpage was about 324,000 acre-ft (450 ft^3/s) and ranged from 306,000 acre-ft (425 ft^3/s) in 1995 to 370,000 acre-ft (510 ft^3/s) in 2001.

The model simulation period was extended by repeating the last 8-year period an additional 3 times, for a total of 66 years (42-year period of the ECM plus the additional 24 years). Streamflow differences are compared for 2018–2025 and each year is compared to the corresponding ECM year for the 8-year period. For example, 2018 is compared to 1994 and 2025 is compared to 2001; thus, pumpage, diversions, prescribed flows, and recharge were the same for each of the corresponding years for the comparison.

The scenario addresses the issue of how the flow system, relative to streamflow, may evolve toward equilibrium under current conditions The simulation does not represent a forecast of the actual groundwater system because of numerous design characteristics implicitly included in this scenario. These characteristics include model inputs that are the same for each 8-year cycle, including: (1) current-condition recharge, (2) irrigation practices, (3) crop types to the extent that there are no changes in streamflow, diversions, and recharge, (4) reservoir operations, (5) diversions, and (6) rates of groundwater withdrawal. Note that the potential impacts of climate change, which can affect most of the above factors—from reservoir operations to diversions to recharge, also are not included in this projection, but the model could be used to analyze potential climate-change effects at some time in the future. The projection period was limited to 24 years because of the uncertainties in the potential effects of climate change on future water supply and demand, as these effects may be large in the basin (Mastin, 2008).

The change in streamflow after 24 years due to existing pumpage varies by location along the river system. As measured by the streamflow at Kiona (the most downstream gaging site in the basin), simulated streamflow decreases from 2018 to 2025 relative to 1994 to 2001 ranged from about 7 to 35 ft^3/s (fig. 35). For the upper basin and Naches,

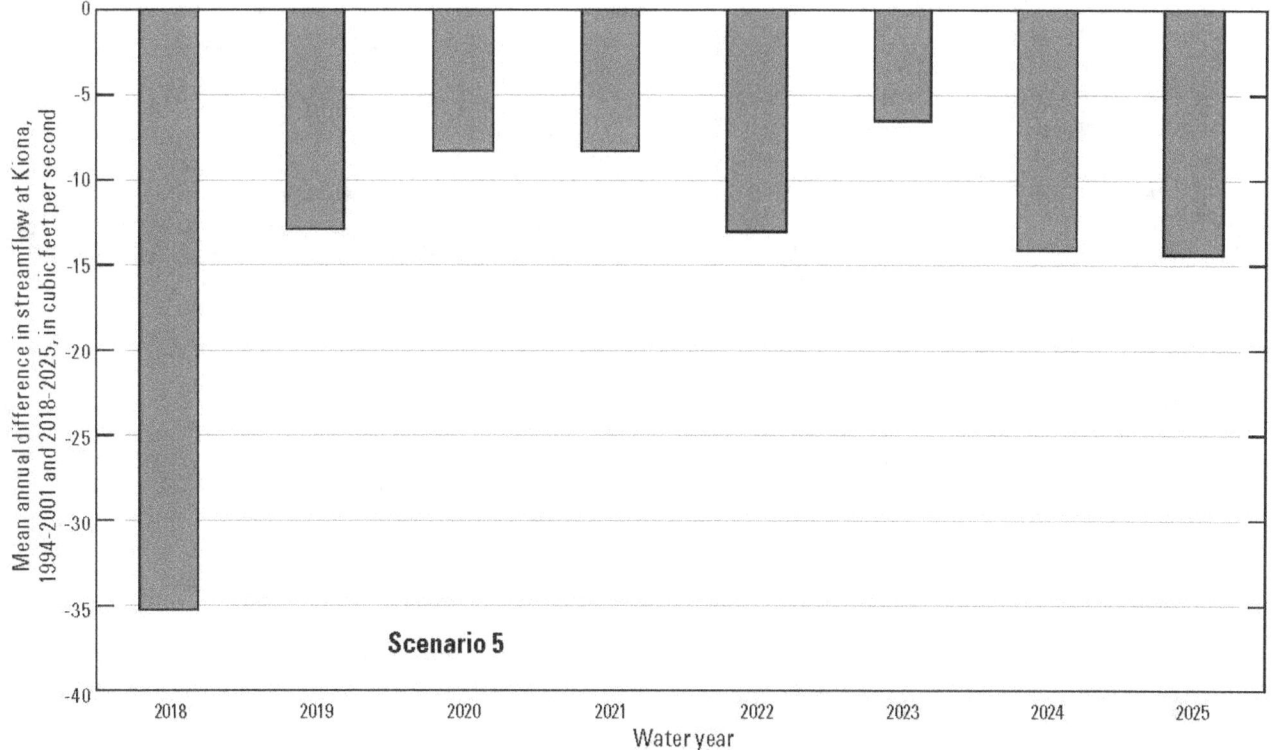

Figure 35.   Differences in streamflow for existing-conditions pumpage continued into the future for 24 years and existing conditions simulated streamflow, Yakima River basin aquifer system, Washington, water years 1994–2001.

the decreases were reasonably uniform over the last 8 years of the simulation and were less than 5 ft³/s. At Union Gap, decreases were slightly larger and ranged from about 2 to 13 ft³/s and increased by about 50 percent at Parker. The change in additional streamflow decreases between Umtanum and Union Gap principally is attributed to pumping in the Selah and Yakima basins. At Mabton, Prosser, and Kiona, the simulated decreases were similar. The largest annual decrease (about 38 ft³/s) was simulated at Richland and the 8-year average decrease was about 5 percent larger than the decrease at Kiona. Streamflow differences increase in a downstream direction because the largest amount of pumpage is from Yakima and Benton Counties, with the amount of pumpage from upstream of Umtanum accounting for less than 8 percent of the total in 2000 (Vaccaro and Sumioka, 2006). Similar to the other scenarios, there is much more variability downstream of Parker due to the complex system responses to basalt pumping. Overall, after 24 years the proportion of pumpage met by storage decreased with a concomitant increase in streamflow decreases. The results suggest that the quantity of streamflow decreases simulated with scenario 1 may increase by 5 percent.

Changes in the Cumulative Water Budgets for Scenarios 1–4

The change in the cumulative water budget for four of the five model applications from the ECM water budget provides a summary of the simulated effects of the scenarios. The future projection (scenario 5) is not included because it is not directly comparable. The recharge component also is not discussed because, except for the exempt well scenario (scenario 3), recharge was held constant for all simulations. To develop easily comparable relative values, all water-budget components for the ECM and scenario simulations were normalized (divided) by the smallest cumulative component for all simulations. The smallest component was the specified-head (CHD) cell inflow to cells from the no-pumpage scenario (scenario 1); CHD-inflow also was the smallest component for the other scenarios. The normalized values from the scenarios for each water-budget component were then subtracted from the ECM values.

The change in water-budget components (fig. 36) shows a direct relation between pumpage quantities leaving the cells and storage changes in the cells. As described previously, the no-pumpage scenario indicated that about 40–45 percent of the pumpage is met by groundwater discharge to streams and the remainder primarily by aquifer storage changes. The combined stream leakage terms, when compared to the pumpage, further indicate that about 40–45 percent is met by streamflow changes. The no-basalt-pumpage scenario (scenario 2) changes highlight that basalt pumpage is being met more by storage changes rather than stream leakage, and that its stream-leakage component is similar to the no-exempt pumpage scenario (scenario 3) stream leakage component. That is, the no-exempt pumpage scenario had a 9 percent decrease in mean annual pumpage compared to the 41 percent decrease for no-basalt pumpage, but the differences in simulated streamflow and stream-leakage component were relatively similar to those from scenario 2. The relatively smaller difference between these scenarios occurs because much of the exempt pumpage is from the basin-fill units that are more connected to the surface-water resources, while pumpage from the basalt units tends to occur deeper in the flow system.

For the pending groundwater application scenario (scenario 4), the graphs (fig. 36) show that much of the application pumpage was met by storage changes over the 8-year period and is consistent with the simulated changes in streamflow from the ECM-simulated values. Eventually, the pumpage met by changes in storage could be met by stream leakage and reduced groundwater discharge as water in storage is depleted.

The degree to which the effects of changes under existing conditions are spread among the different aggregated units (basin-fill units and basalt units) will continue to change over time as the groundwater-flow system equilibrates to existing conditions. Any changes in pumpage, recharge, diversions, and reservoir releases will influence these relations and the time it takes for the system to approach equilibrium conditions.

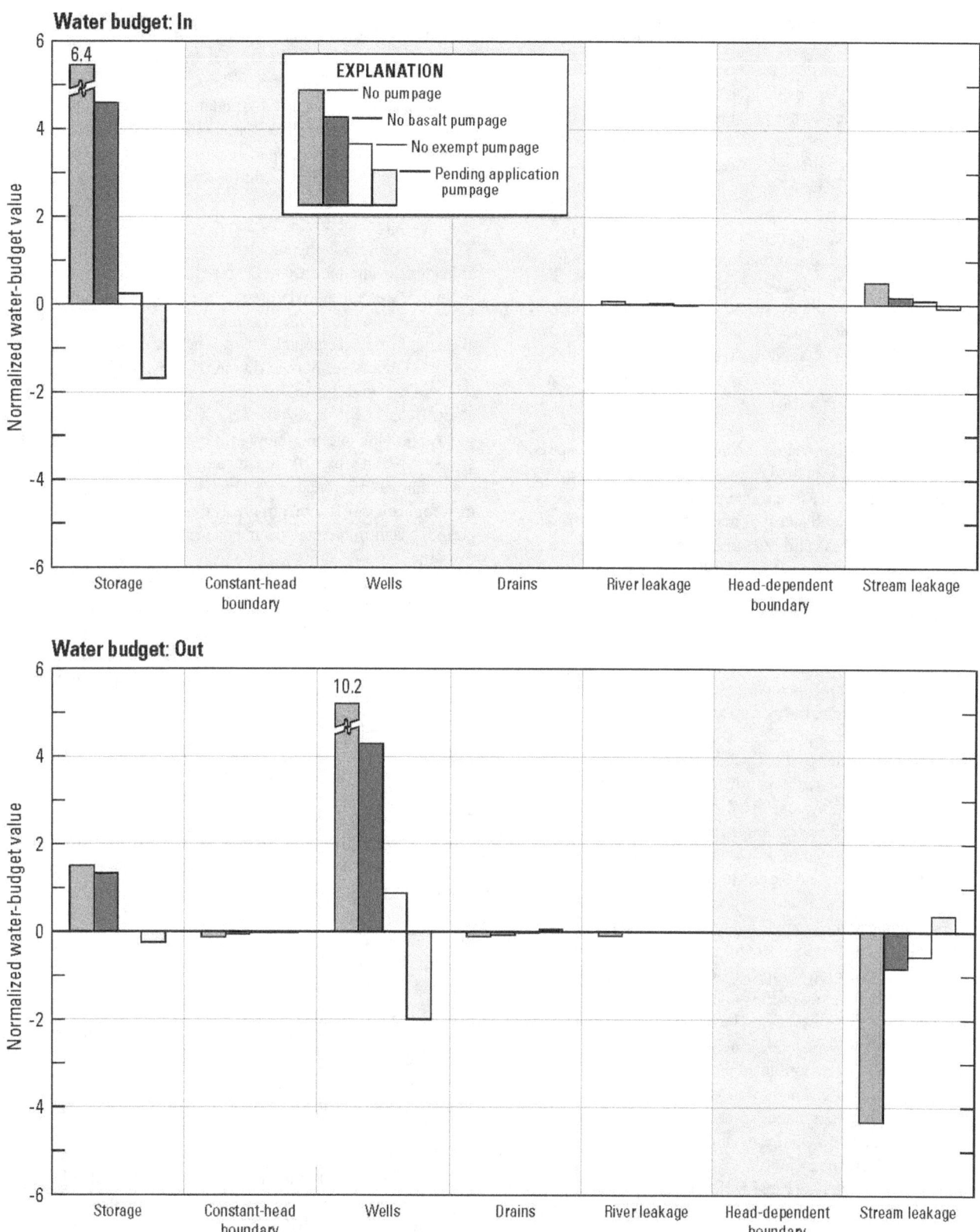

Figure 36. Differences in the normalized, cumulative groundwater budget components between existing-conditions model and four model scenarios, Yakima River basin aquifer system, Washington.

Summary and Conclusions

A regional, transient numerical model of groundwater flow was constructed for the Yakima River basin aquifer system to better understand the groundwater-flow system and its relation to surface-water resources. The existing-conditions model (ECM) described in this report can be used as a tool for water-management agencies. The ECM was constructed using the U.S. Geological Survey finite-difference model MODFLOW-2000, and integrates information from all previously completed and documented study components. The ECM uses 1,000-foot grid cells that subdivide the model domain into 600 rows and 600 columns. Forty-eight hydrogeologic units (HGUs) in the model are included in 24 model layers. The Yakima River, all major tributaries, and major agricultural drains are included as either drain or streamflow routing cells. Recharge was estimated from previous work using physical-process models. Groundwater pumpage is based on values estimated in a previous study component; values include estimates for most wells with estimated standby/reserve rights that are used in drought years. The temporal discretization of the ECM is a 1-day time step.

The ECM was calibrated to the transient conditions for the 42-year period from October 1959 to September 2001, resulting in 504 1-month stress periods. Calibration was completed by using both traditional trial-and-error methods and automated parameter-estimation techniques. The model simulates the shape and slope of the water table and generally is consistent with mapped water levels. At 2,196 well measurement points, the average difference between simulated and measured hydraulic heads is -49 feet and the root-mean-square error divided by the total difference in water levels is 4 percent. Simulated river streamflow was compared to measured streamflow at selected sites. The percent differences between measured and simulated mean annual flows for the 42-year period ranged from less than 1 to about 9 percent for Yakima River sites and 5 percent for the Naches River. For all Yakima River sites and the Naches River, 58 percent of the simulated annual values were within 10 percent of the measured values, and 63 percent were within 15 percent.

Five applications of the model were completed to obtain a better understanding of the relation between groundwater and surface-water resources. The model applications were simulated as five scenarios, which were then used to quantify the relation between simulated pumpage and streamflow. For three applications or scenarios, the transient model was operated with no pumpage (scenario 1), no basalt pumpage (scenario 2), and no exempt-well pumpage (scenario 3).

The difference in simulated streamflow between scenario 1 and the calibrated ECM increased over the 42-year period and followed the trend in the difference in pumpage, and the difference also increased in a downstream direction. The annual difference in streamflow at the Yakima River at Umtanum (Umtanum) was computed to be 18 cubic feet per second (ft^3/s) in 2001, whereas the annual difference at the Yakima River at Richland (Richland) in 2001 was 194 ft^3/s. This difference did not linearly increase in a downstream direction because of tributary inflows, diversions, and returns. As a percentage of the ECM-simulated annual streamflow, the mean annual difference in streamflow at Richland was 5 percent (about 45 percent of the mean annual pumpage), and the mean annual difference for the sites ranged from less than 1 percent at the Yakima River at Horlick (Horlick) to as large as about 6 percent at the Yakima River at Prosser (Prosser).

For scenario 2, the simulated effect of basalt pumpage indicates the annual difference in streamflow, as measured at Richland, was estimated to range from nearly zero in the early 1960s to 57 ft^3/s in 1998. The largest simulated annual change at Umtanum was less than 3 ft^3/s, whereas the largest changes at the Yakima River above Ahtanum Creek, at Union Gap (Union Gap) and the Yakima River near Parker (Parker) were about 14 and 29 ft^3/s, respectively. The mean annual difference as a percentage of streamflow at Richland was about 1 percent or about 19 percent of the mean annual basalt pumpage. The mean annual difference at Richland was about 134 ft^3/s lower than for scenario 1, indicating that during the 42-year ECM period, basalt pumpage was not as important as pumpage from the basin-fill units relative to simulated effects on the surface-water resources. That is, basalt pumpage accounted for about 16 to 17 percent of the mean annual change. For all years, the change due to basalt pumpage ranged from about 0 to 34 percent of the total change simulated in scenario 1.

For scenario 3, the annual differences between the ECM with the addition of recharge from septic drainage and the component of this scenario without exempt pumpage were small compared to scenario 1. The differences between calibrated ECM and scenario 3 simulated streamflow were less than 1 ft^3/s at Horlick, and at Umtanum and the Naches River at North Yakima (Naches) differences were less than 2 ft^3/s. Annual differences in streamflow at Parker were larger by about 2–4 ft^3/s. Thus, the simulated effects of exempt pumpage on the Yakima River were less than for scenarios 1 and 2, and for the lower 17 miles of the Naches River, the effects also were small. During model calibration, changes in streamflow from exempt pumpage were sensitive to basin-fill parameter adjustments.

For scenario 4, the ECM was operated from 1994 through 2001 under existing conditions, but with additional pumpage estimated for pending groundwater applications. The differences in streamflow from the calibrated model and this scenario indicated decreases in streamflow. The simulated potential effects of the estimated application pumpage on streamflow compound those estimated in scenario 1. After 8 years, effects at Umtanum and Naches were relatively small (about 14 and 7 ft^3/s, respectively). The additional pumpage from the applications led to a 37 ft^3/s decrease in streamflow at Parker, and by Prosser a 73 ft^3/s decrease, which was similar to the decrease at Mabton and Kiona. In 2001, streamflow was decreased about 88 ft³/s at Richland. The decrease at Richland is in addition to the simulated effect of 194 ft³/s from all pumpage in 2001. Between about 6–29 percent of the estimated annual pending application pumpage (330 ft³/s) was met by decreased streamflow in the modeled area. Although the estimated application pumpage was about 73 percent of the ECM 8-year average pumpage, the percentage of pumpage met by decreased streamflow was not as large as in scenario 1 because much of the application pumpage occurs from the basalt units that are relatively more isolated from surface-water resources. The change in streamflow would increase over time as the system response evolves toward equilibrium.

For scenario 5, the existing conditions for 1994 through 2001 were projected through 2025. The scenario only estimates the potential effects of existing pumping if it continued into the future without any other changes, and it does not represent a forecast of the actual groundwater system because of numerous factors that are implicitly included in this scenario. The change in streamflow after 24 years due to existing pumping varied by location along the river system. As measured by the streamflow at Kiona (the most downstream gaging site in the basin), simulated streamflow decreases from 2018 to 2025 relative to 1994 to 2001 ranged between about 7 and 35 ft³/s. These decreases would be in addition to the simulated decreases estimated from scenario 1 from 1994 to 2001. For the upper basin and Naches, the decreases were reasonably uniform over the last 8 years of the simulation and were less than 5 ft³/s. At Union Gap, decreases were slightly larger and ranged from about 2 to 13 ft³/s and increased by about 50 percent at Parker. The change in additional streamflow decreases between Umtanum and Union Gap principally is attributed to pumping in the Selah and Yakima basins. At Mabton, Prosser, and Kiona, the simulated decreases were very similar. The largest annual decrease (about 38 ft³/s) was simulated at Richland. The effect of pumping on streamflow increases in a downstream direction because the amount of pumpage is largest from Yakima and Benton Counties, with the amount of pumpage from upstream of Umtanum accounting for less than 8 percent of the total in 2000. Similar to the other scenarios, there is much more variability downstream of Parker due to the complex system responses to basalt pumping. For the modeled region as a whole, as measured by the simulated streamflow at Richland, the 8-year average decrease in streamflow for scenario 1 represented about 46 percent of the 8-year mean annual pumpage compared to a reduction of 52 percent for scenario 5.

In summary, a complex and comprehensive transient groundwater-flow model was constructed for the Yakima River basin aquifer system. The model simulates streamflow for much of the Yakima River and the lower 17 miles of the Naches River. The model was calibrated to both hydraulic heads and streamflow, and the differences between measured and simulated values were within acceptable limits. The model reasonably represents the groundwater-flow system as compared to the flow system mapped and described as part of this study. The five model-application scenarios indicated potential streamflow differences due to changes to the existing pumpage from 1960 through 2001. The quantity of streamflow differences generally was directly related to the total pumpage in a model scenario. The scenarios indicate the potential usefulness of the model for assessing water-management strategies in such a large and complex system.

Acknowledgments

The authors wish to thank the government agencies and groups such as the Bureau of Reclamation, Washington State Department of Ecology, the Yakama Nation, Bureau of Indian Affairs, conservation districts, and irrigation districts, that provided valuable support, access to their records, and data during this study.

The authors also would like to thank the study team members from the cooperating entities that contributed to the study through numerous technical meetings and discussions throughout the study. The study team members were: Kayti Didricksen with the Bureau of Reclamation, John Kirk with the Washington State Department of Ecology, Tom Ring with the Yakama Nation, and Frank Spane with Pacific Northwest National Laboratory.

References Cited

Anderman, E.R., and Hill, M.C., 2000, MODFLOW-2000, the U.S. Geological Survey Modular Ground-Water Model -Documentation of the Hydrogeologic-Unit Flow (HUF) Package: U.S. Geological Survey Open-File Report 2000-342, 89 p. (Also available at http://pubs.er.usgs.gov/usgspubs/ofr/ofr00342.)

Anderson, M.R., and Woessner, W.W., 1992, Applied groundwater modeling simulation of flow and advective transport: San Diego/New York/Boston/London/Sydney/Tokyo/Toronto, Academic Press, Inc., 381 p.

Bauer, H.H., and Hansen, A.J., Jr., 2000, Hydrology of the Columbia Plateau regional aquifer system, Washington, Oregon, and Idaho: U.S. Geological Survey Water-Resources Investigations Report 96-4106, 61 p. (Also available at http://pubs.er.usgs.gov/usgspubs/wri/wri964106.)

Bauer, H.H., and Vaccaro, J.J., 1987, Documentation of a deep percolation model for estimating ground-water recharge: U.S. Geological Survey Open-File Report 86-536, 180 p.

Bureau of Reclamation, 1999, Yakima River Basin Water Enhancement Project, Washington, final programmatic environmental impact statement: U.S. Department of Interior, Bureau of Reclamation, Pacific Northwest Region, Upper Columbia Area Office, Yakima, Wash., 197 p.

Campbell, N.P., 1989, Structural and stratigraphic interpretation of rocks under the Yakima fold belt, Columbia Basin, based on recent surface mapping and well data, *in* Reidel, S.P., and Hooper, P.R., eds., Volcanism and tectonism in the Columbia River flood-basalt province: Boulder, Colo., Geological Society of America, Special Paper 239, p. 209-222.

Cline, D.R., and Collins, C.A., 1993, Ground-water pumpage in the Columbia Plateau, Washington and Oregon, 1945 to 1984: U.S. Geological Survey Water-Resources Investigations Report 90-4085, 31 p., 5 pls.

Cline, D.R., and Collins, C.A., 1992, Ground-water pumpage in the Columbia Plateau, Washington and Oregon, 1945-84, *in*, Prince, K.R., and Johnson, A.I., eds., Aquifers of the Far West: American Water Resources Association, Monograph Series no. 16, p. 99-107.

Cline, D.R., and Knadle, M.E., 1990, Ground-water pumpage from the Columbia Plateau regional aquifer system, Washington, 1984: U.S. Geological Survey Water-Resources Investigations Report 87-4135, 32 p., 1 sheet. (Also available at http://pubs.er.usgs.gov/publication/wri874135.)

Collins, C.A., 1987, Ground-water pumpage from the Columbia Plateau regional aquifer system, Oregon, 1984: U.S. Geological Survey Water-Resources Investigations Report 86-4211, 21 p.

Cuffney, T.F., Meador, M.R., Porter, S.D., and Gurtz, M.E., 1997, Distribution of fish, benthic invertebrate, and algal communities in relation to physical and chemical conditions, Yakima River Basin, Washington, 1990: U.S. Geological Survey Water-Resources Investigations Report 96-4280, 94 p. (Also available at http://or.water.usgs.gov/pubs_dir/Abstracts/96-4280.html.)

Davies-Smith, A., Bolke, E.L., and Collins, C.A., 1989, Geohydrology and digital simulation of the ground-water flow system in the Umatilla Plateau and Horse Heaven Hills area, Oregon and Washington: U.S. Geological Survey Water-Resources Investigations Report 87-4268, 72 p. (Also available at http://pubs.er.usgs.gov/usgspubs/wri/wri874268.)

Doherty, J.E., 2010, PEST, Model-independent parameter estimation—User manual (5th ed., with additions): Brisbane, Australia, Watermark Numerical Computing.

Doherty, J.E., and Hunt, R.J., 2010, Approaches to highly parameterized inversion—A guide to using PEST for groundwater-model calibration: U.S. Geological Survey Scientific Investigations Report 2010–5169, 59 p. (Also available at http://pubs.usgs.gov/sir/2010/5169/.)

Drost, B.W., Whiteman, K.J., and Gonthier, J.B., 1990, The geologic framework of the Columbia Plateau regional aquifer system, Washington, Oregon, and Idaho: U.S. Geological Survey Water-Resources Investigations Report 87-4238, 10 p., 10 sheets. (Also available at http://pubs.er.usgs.gov/usgspubs/wri/wri874238.)

Drost, B.W., Ely, D.M., and Lum, W.E., 1999, Conceptual model and numerical simulation of the ground-water flow system in the unconsolidated sediments of Thurston County, Washington: U.S. Geological Survey Water-Resources Investigations Report 99-4165, 106 p. (Also available at http://pubs.usgs.gov/wri/wri994165/.)

Faunt, C.C., Blainey, J.B., Hill, M.C., D'Agnese, F.A., and O'Brien, G.M., 2004, Death Valley regional ground-water flow system, Nevada and California—Hydrogeologic framework and transient ground-water flow model, ed., Belcher, W.R.: U.S. Geological Survey Scientific Investigations Report 2004-5205, Chapter F, p. 265-362. (Also available at http://pubs.er.usgs.gov/usgspubs/sir/sir20045205.)

Faunt, C.C., Hanson, R.T., Belitz, K., Schmid, W. , Predmore, S.P., Rewis, D.L., and McPherson, K., 2009, Numerical model of the hydrologic landscape and groundwater flow in California's Central Valley: U.S. Geological Survey Professional Paper 1766, Chapter C, p. 121-225. (Also available at http://pubs.er.usgs.gov/usgspubs/pp/pp1766.)

Flaherty, N.M., 1975, The Yakima Basin and its water: Washington State University, Water Resources Center, Pullman, Wash., 29 p.

Franke, O.L., Reilly, T.E., and Bennett, G.D, 1987, Definition of boundary and initial conditions in the analysis of saturated ground-water flow systems-An introduction: U.S. Geological Survey Techniques of Water-Resources Investigations, Book 3, Applications of Hydraulics, Chapter B5, 15 p.

Fuhrer, G.J., McKenzie, S.W., Rinella, J.F., and Skach, K.A., 1994, Surface-water assessment of the Yakima River basin in Washington—Analysis of major and minor elements in fine-grained-streambed sediment, 1987.: U.S. Geological Survey Open-File Report 93-30, 226 p., accessed January 13, 2009(Also available at http://pubs.er.usgs.gov/usgspubs/ofr/ofr9330.)

Glantz, M.H., 1982, Consequences and responsibilities in drought forecasting—The case of Yakima, 1977: Water Resources Research, v. 18, no. 1, p. 3-13.

Gulick, C.W., and Korosec, M.A., 1990, Geologic map of the Banks Lake 1:100,000 quadrangle, Washington: Washington Division of Geology and Earth Resources, Open-File Report 90-6, 20 p., 1 sheet, scale 1:100,000.

Hansen, A.J., Vaccaro, J.J., and Bauer, H.H., 1994, Ground-water flow simulation of the Columbia Plateau Regional Aquifer System, Washington, Oregon, and Idaho: U.S. Geological Survey Water-Resources Investigations Report 91-4187, 81 p., 15 pls.

Harbaugh, A.W., Banta, E.R., Hill, M.C., and McDonald, M.G., 2000, MODFLOW-2000, the U.S. Geological Survey modular ground-water model—user guide to modularization concepts and the ground-water flow process: U.S. Geological Survey Open-File Report 00-92, 121 p. (Also available at http://pubs.er.usgs.gov/publication/ofr200092.)

Hill, M.C., 1998, Methods and Guidelines for effective model calibration: U.S. Geological Survey Water-Resources Investigations Report 98-4005, 90 p. (Also available at http://pubs.er.usgs.gov/#search:basic/query=Methods%20 and%20Guidelines%20for%20effective%20model%20 calibration/page=1/page_size=100:0.)

Hill, M.C., and Tiedeman, C.R, 2003, Weighting observations in the context of calibrating ground-water models, *in* Kovar, K., and Zbynek, H., eds., Calibration and reliability in groundwater modeling, a few steps closer to reality: International Association of Hydrological Sciences Publication, v. 277, p. 196–203.

Hsieh, P.A., and Freckleton, J.R., 1993, Documentation of a computer program to simulate horizontal-flow barriers using the U.S. Geological Survey's modular three-dimensional finite-difference ground-water flow model: U.S. Geological Survey Open-File Report 92-477, 32 p. (Also available at http://pubs.er.usgs.gov/djvu/OFR/1992/ofr_92_477.djvu.)

Jones, M.A., Vaccaro, J.J., and Watkins, A.M., 2006, Hydrogeologic framework of sedimentary deposits in six structural basins, Yakima River Basin, Washington: U.S. Geological Survey Scientific Investigations Report 2006-5116, 24 p., 7 pls. (Also available at http://pubs.water.usgs.gov/sir2006-5116.)

Jones, M.A., and Vaccaro, J.J., 2008, Extent and depth to top of basalt and interbed hydrogeologic units, Yakima River Basin Aquifer System, Washington: U.S. Geological Survey Scientific Investigations Report 2008–5045, 22 p., 5 pls. (Also available at http://pubs.usgs.gov/sir/2008/5045/.)

Keys, M.E., Vaccaro, J.J., Jones, M.A., and Julich, R.J., 2008, Hydrographs showing ground-water level trends for selected wells in the Yakima River Basin aquifer system, Washington: U.S. Geological Survey Data Series 343. (Also available at http://pubs.usgs.gov/ds/343/.)

Kinnison, H.B., and Sceva, J.E., 1963, Effects of hydraulic and geologic factors on streamflow of the Yakima River Basin, Washington: U.S. Geological Survey Water-Supply Paper 1595, 134 p. (Also available at http://pubs.er.usgs.gov/usgspubs/wsp/wsp1595.)

Kirk, T.K. and Mackie, T.L., 1993, Black Rock-Moxee Valley groundwater study: Washington State Department of Ecology, Open File Technical Report 93-1, 79 p., accessed August 16, 2011 at http://www.ecy.wa.gov/biblio/oftr9301.html.

Konikow, L.F., Hornberger, G.Z., Halford, K.J., and Hanson, R.T., 2009, Revised multi-node well (MNW2) package for MODFLOW ground-water flow model: U.S. Geological Survey Techniques and Methods 6–A30, 67 p.

Korosec, M.A., 1987, Geologic map of the Mount Adams quadrangle, Washington: Washington State Department of Natural Resources, Open-File Report 87-5, 41 p., 1 sheet, scale 1:100,000.

Kratz, M.R., 1978, Dilemmas, disruptions, but no disaster—Drought in the Yakima basin, Washington, 1977: State Climatologist for Arizona Climatological Publications, Scientific Paper, no. 3, 16 p.

Leavesley, G.H., Lichty, R.W., Troutman, B.M., and Saindon, L.G., 1983, Precipitation-runoff modeling system-User's manual: U.S. Geological Survey Water-Resources Investigations Report 83-4238, 207 p. (Also available at http://pubs.er.usgs.gov/djvu/WRI/wrir_83_4238.djvu.)

Leavesley, G.H., Restrepo, P.J., Markstrom, S.L., Dixon, M., and Stannard, L.G., 1996, The modular modeling system (MMS)-User's manual: U.S. Geological Survey Open-File Report 96-151, 200 p.

Lindolm, G.F., and Vaccaro, J.J., 1988, Region 2, Columbia Lava Plateau, in Back, William, Rosenshein, J.S., and Seaber, P.R., eds., Hydrogeology, v. O-2—The geology of North America: Boulder, Colo., Geological Society of America, p. 37-50.

Long, P.E., and Wood, B.J., 1986, Structures, textures, and cooling histories of Columbia River Basalt flows: Geological Society of America Bulletin, v. 97, no. 9, p. 1,144-1,155.

Lum, W.E., II, Smoot, J.L., and Ralston, D.R., 1990, Geohydrology and numerical model analysis of ground-water flow in the Pullman-Moscow area, Washington and Idaho: U.S. Geological Survey Water-Resources Investigations Report 89-4103, 73 p. (Also available at http://pubs.er.usgs.gov/usgspubs/wri/wri894103.)

MacDonald, G.A., 1967, Forms and structures of extrusive basaltic rocks, *in* Hess, H.H. and Poldervaart, A., eds., Basalts—The Poldervaart treatise on rocks of basaltic composition, v. 1: New York, Wiley and Sons, p. 482.

Magirl, C.S., Julich, R.J., Welch, W.B., Curran, C.R., Mastin, M.C., and Vaccaro, J.J., 2009, Summary of seepage investigations in the Yakima River basin, Washington: U.S. Geological Survey Data Series 473. (Also available at http://pubs.usgs.gov/ds/473.)

Magirl, C.S., and Olsen, T.D., 2009, Navigability potential of Washington rivers and streams determined with hydraulic geometry and a geographic information system: U.S. Geological Survey Scientific Investigations Report 2009–5122, 22 p. (Also available at http://pubs.usgs.gov/sir/2009/5122/.)

Mastin, M.C., 2008, Effects of potential future warming on runoff in the Yakima River Basin, Washington: U.S. Geological Survey Scientific Investigations Report 2008-5124, 12 p. (Also available at http://pubs.usgs.gov/sir/2008/5124.)

Mastin, M.C., and Vaccaro, J.J., 2002, Watershed models for decision support in the Yakima River Basin, Washington: U.S. Geological Survey Open-File Report 02-404, 46 p. (Also available at http://pubs.usgs.gov/of/2002/ofr02404.)

Meyers, C.W., and Price, S.M., 1979, Geologic studies of the Columbia Plateau, a status report: Rockwell International, Rockwell Hanford Operations RHO-BWI-ST-4, 520 p.

Newcomb, R.C., 1965, Geology and ground-water resources of the Walla Walla River basin, Washington-Oregon: Washington Division of Water Resources, Water-Supply Bulletin 21, 151 p., accessed January 31, 2011 at http://www.ecy.wa.gov/programs/eap/wsb/pdfs/WSB_21_Book.pdf.

Newcomb, R.C., 1969, Effect of tectonic structure on the occurrence of ground water in the basalt of the Columbia River Group of The Dalles area, Oregon and Washington: U.S. Geological Survey Professional Paper 383-C, 33 p., accessed August 16, 2011 at http://pubs.er.usgs.gov/djvu/PP/pp_383_c.djvu.

Niswonger, R.G. and Prudic, D.E., 2005, Documentation of the Streamflow-Routing (SFR2) Package to include unsaturated flow beneath streams–A modification to SFR1: U.S. Geological Survey Techniques and Methods, Book 6, Chap. A13, 47 p. (Also available at http://pubs.usgs.gov/tm/2006/tm6A13/.)

Omernik, J.M., 1987, Ecoregions of the conterminous United States: Annals of the Association of American Geographers, v. 77, no. 1, p. 118-125.

Packard, F.A, Hansen, A.J., Jr., and Bauer, H.H., 1996, Hydrogeology and simulation of flow and the effects of development alternatives on the basalt aquifers of the Horse Heaven Hills, south-central Washington: U.S. Geological Survey Water-Resources Investigations Report 94-4068, 92 p., 2 plates,(Also available at http://pubs.er.usgs.gov/usgspubs/wri/wri944068.)

Parker, G.L, and Storey, F.B, 1916, Water powers of the Cascade Range, Part III, Yakima River Basin: U.S. Geological Survey Water-Supply Paper 369, 169 p., 18 pls.

Phillips, W.M., and Walsh, T.J., 1987, Geologic map of the northwest part of the Goldendale quadrangle, Washington: Washington Division of Geology and Earth Resources, Open-File Report 87-13, 9 p., 1 sheet, scale 1:100,000.

Poeter, E.P., and Hill, M.C., 1998, Documentation of UCODE, a computer code for universal inverse modeling: U.S. Geological Survey Water-Resources Investigations Report 98-4080, 116 p.

Reidel, S.P., 1982, Stratigraphy of the Grande Ronde Basalt, Columbia River Basalt Group, from the lower Salmon River and northern Hells Canyon area, Idaho, Oregon, and Washington, in Bonnichsen, B., and Breckenridge, R.M., eds., Cenozoic geology of Idaho: Idaho Bureau of Mines and Geology, Bulletin 26, p. 77-101.

Reidel, S.P., and Fecht, K.R., 1994a, Geologic map of the Priest Rapids 1:100,000 quadrangle, Washington: Washington Division of Geology and Earth Resources, Open-File Report 94-13, 22 p., scale 1:100,000.

Reidel, S.P., and Fecht, K.R., 1994b, Geologic map of the Richland 1:100,000 quadrangle, Washington: Washington Division of Geology and Earth Resources, Open-File Report 94-8, 21 p.

Reidel, S.P., Johnson, V.G., and Spane, F.A., 2002, Natural gas storage in basalt aquifers of the Columbia Basin, Pacific Northwest USA—A guide to site characterization: Pacific Northwest National Laboratory, Richland, Wash., variously paginated, accessed August 16, 2011 at http://www.pnl.gov/main/publications/external/technical_reports/PNNL-13962.pdf.

Reilly, T.E., and Harbaugh, A.W., 2004, Guidelines for evaluating ground-water flow models: U.S. Geological Survey Scientific Investigations Report 2004-5038, 30 p. (Also available at http://pubs.usgs.gov/sir/2004/5038/.)

Schasse, H.W., 1987, Geologic map of the Mount Rainier quadrangle, Washington: Washington State Department of Natural Resources, Open-File Report 87-16, 43 p., 1 sheet, scale 1:100,000.

Schuster, J.E., 1994a, Geologic maps of the east half of the Washington portion of the Goldendale 1:100,000 quadrangle and the Washington portion of the Hermiston 1:100,000 quadrangle: Washington State Department of Natural Resources, Open-File Report 94-9, 17 p., 1 sheet.

Schuster, J.E., 1994b, Geologic map of the east half of the Toppenish 1:100,000 quadrangle, Washington: Washington State Department of Natural Resources, Open-File Report 94-10, 15 p., 1 sheet.

Schuster, J.E., 1994c, Geologic map of the east half of the Yakima 1:100,000 quadrangle, Washington: Washington State Department of Natural Resources, Open-File Report 94-12, 19 p., 1 sheet.

Stearns, R.T., 1942, Hydrology of lava-rock terranes, *in* Meinzer, O.E., ed., Hydrology: New York, McGraw-Hill, chap. WV, 712 p.

Sublette, W.R, 1986, Rock mechanics data package, Rev. 1: Richland, Washington, Rockwell Hanford Operations Report SD- BWI-DP-041, 78 p.

Swanson, D.A., Anderson, J.L., Bentley, R.D., Byerly, G.R., Camp, V.E., Gardner, J.N., and Wright, T.L., 1979a, Reconnaissance geologic map of the Columbia River Basalt Group in eastern Washington and northern Idaho: U.S. Geological Survey Open-File Report 79-1363, 26 p., 12 sheets, scale 1:250,000.

Swanson, D.A., Brown, J.C., Anderson, J.L., Bentley, R.D., Byerly, G.R., Gardner, J.N., and Wright, T.L., 1979b, Preliminary structure contour maps on the top of the Grande Ronde and Wanapum Basalts, eastern Washington and northern Idaho: U.S. Geological Survey Open-File Report 79-1364, 3 sheets.

Swanson, D.A., Wright, T.L., Hooper, T.R., and Bentley, R.D., 1979c, Revisions in stratigraphic nomenclature of the Columbia River Basalt Group: U.S. Geological Survey Bulletin 1457-G, 59 p. (Also available at http://pubs.er.usgs.gov/usgspubs/b/b1457G.)

Swanson, D.A., and Wright, T.L., 1978, Bedrock geology of the southern Columbia Plateau and adjacent areas, Chap. 3, *in* Baker, V.R., and Nummedal, D., eds., The channeled scabland: Washington, D.C., Planetary Geology Program, National Aeronautical and Space Administration, p. 37-57.

Systems Operations Advisory Committee, 1999, Report on biologically based flows for the Yakima River basin: The Systems Operations Advisory Committee, Report to The Secretary of the Interior, Yakima, Washington, 58 p., 1 appendix.

Tabor, R.W., Frizzell, V.A., Jr., Booth, D.B., Waitt, R.B., Whetten, J.T., and Zartman, R.E., 1993, Geologic map of the Skykomish River 30- by 60-minute quadrangle, Washington: U.S. Geological Survey Miscellaneous Investigations Series Map I-1963, 42 p., 1 sheet, scale 1:100,000.

Tabor, R.W., Frizzell, V.A., Jr., Whetten, J.T., Waitt, R.B., Swanson, D.A., Byerly, G.R., Booth, D.B., Hetherington, M.J., and Zartman, R.E., 1987, Geologic map of the Chelan 30-minute by 60-minute quadrangle, Washington: U.S. Geological Survey Miscellaneous Investigations Series Map I-1661, 33 p., 1 sheet, scale 1:100,000.

Tabor, R.W., Waitt, R.B., Jr., Frizzell, V.A., Jr., Swanson, D.A., Byerly, G.R., and Bentley, R.D., 1982, Geologic map of the Wenatchee 1:100,000 quadrangle, central Washington: U.S. Geological Survey Miscellaneous Investigations Series Map I-1311, 26 p., 1 sheet.

Tanaka, H.H., Barrett, G.T., and Wildrick, L., 1979, Regional basalt hydrology of the Columbia Plateau in Washington: Richland, Washington, Rockwell Hanford Operations, RHO-BWI-C-60, 303 p.

Theis, C.V., 1940, The source of water derived from wells— Essential factors controlling the response of an aquifer to development: Civil Engineering, v. 10, p. 277-280.

Tiedeman, C.R., Hill, M.C., D'Agnese, F.A., and Faunt, C.C., 2003, Methods for using ground-water model predictions to guide hydrogeologic data collection with application to the Death Valley regional groundwater flow system: Water Resources Research, v. 39, no. 1, p. 5–1 to 5–17.

Tolan, T.L., Reidel, S.P., Beeson, M.H., Anderson, J.L., Fecht, K.R., and Swanson, D.A., 1989, Revisions to the estimates of the areal extent and volume of the Columbia River Basalt Group, *in* Reidel, S.P., and Hooper, P.R., eds., Volcanism and tectonism in the Columbia River flood-basalt province: Boulder, Colorado, Geological Society of America, Special Paper 239, p. 1-20.

Tomkeieff, S.I., 1940, The basalt lavas of the Giant's Causeway district of Northern Ireland: Bulletin Volcanologique, v. 6, p. 89–143.

Troutman, B.M., 1985, Errors and parameter estimation in precipitation-runoff modeling—2. Case Study: Water Resources Research, v. 21, no. 8, p. 1,214-1,222.

U.S. District Court, 1945, Consent decree in the District Court of the United States for the Eastern District of Washington, Southern Division: Spokane, Washington, Civil action, no. 21.

Vaccaro, J.J., 1995, Changes in the hydrometeorological regime in the Pacific Northwest: Proceedings of the Twelfth Annual Pacific Climate Workshop, Asilomar, Cal., Technical Report 46 of the Interagency Ecological Program for the Sacramento-San Joaquin Estuary, p. 143.

Vaccaro, J.J., 2002, Interdecadal changes in the hydrometeorological regime of the Pacific Northwest and in the regional-to-hemispheric climate regimes, and their linkages, 2002: U.S. Geological Survey Water-Resources Investigations Report 02-4176, 94 p. (Also available at http://pubs.er.usgs.gov/usgspubs/wri/wri024176.)

Vaccaro, J.J., 2007, A deep percolation model for estimating ground-water recharge—Documentation of modules for the modular modeling system of the U.S. Geological Survey: U.S. Geological Survey Scientific Investigations Report 2006-5318, 30 p. (Also available at http://pubs.er.usgs.gov/usgspubs/sir/sir20065318.)

Vaccaro, J.J., and Maloy, K.J., 2006, A method to thermally profile long river reaches to identify potential areas of ground-water discharge and preferred salmonid habitat: U.S. Geological Survey Scientific Investigations Report 2006-5136, 16 p. (Also available at http://pubs.usgs.gov/sir/20065136/.)

Vaccaro, J.J., and Olsen, T.D., 2007a, Estimates of ground-water recharge to the Yakima River Basin Aquifer System, Washington, for predevelopment and current land-use and land-cover conditions: U.S. Geological Survey Scientific Investigations Report 2007-5007, 30 p. (Also available at http://pubs.water.usgs.gov/sir2007/5007.)

Vaccaro, J.J., and Olsen, T.D., 2007b, Estimates of monthly ground-water recharge to the Yakima Basin Aquifer System, Washington, 1960-2001, for current land-use and land-cover conditions: U.S. Geological Survey Open File Report 2007-1238, 2 p. (Also available at http://pubs.water.usgs.gov/ofr20071238.)

Vaccaro, J.J., and Sumioka, S.S., 2006, Estimates of ground-water pumpage from the Yakima River Basin aquifer system, Washington, 1960-2000: U.S. Geological Survey Scientific Investigations Report 2006–5205, 56 p. (Also available at http://pubs.usgs.gov/sir/2006/5205.)

Vaccaro, J.J., Keys, M.E., Julich, R.J., and Welch, W.B., 2008, Thermal profiles for selected river reaches in the Yakima River Basin, Washington: U.S. Geological Survey Data Series 342. (Also available at http://pubs.usgs.gov/ds/342.)

Vaccaro, J.J., Jones, M.A., Ely, D.M., Keys, M.E., Olsen, T.D., Welch, W.B., and Cox, S.E., 2009, Hydrogeologic framework of the Yakima River basin aquifer system, Washington: U.S. Geological Survey Scientific Investigations Report 2009–5152, 106 p. (Also available at http://pubs.er.usgs.gov/usgspubs/sir/sir20095152.)

Walsh, T.J., 1986a, Geologic map of the west half of the Toppenish quadrangle, Washington: Washington Division of Geology and Earth Resources, Open-File Report 86-3, 7 p., 1 sheet, scale 1:100,000.

Walsh, T.J., 1986b, Geologic map of the west half of the Yakima quadrangle, Washington: Washington Division of Geology and Earth Resources, Open-File Report 86-4, 9 p., 1 sheet, scale 1:100,000.

Washington State Department of Ecology, 1998, Washington State water law, a primer: WR 98 152, accessed August 16, 2011 at http://www.ecy.wa.gov/pubs/98152.

Washington State Department of Natural Resources, 2002, Digital geology of Washington state: Washington State Department of Natural Resources, 1:100,000 scale quadrangles, accessed April 2002 at url.

Waters, A.C., 1960, Determining direction of flow in basalts: American Journal of Science, v. 258, p. 350-366.

Weiss, Emanuel, 1982, A computer program for calculating relative-transmissivity input arrays to aid model calibration: U.S. Geological Survey Open-File Report 82-447, 18 p.

Winter, T.C., Harvey, J.W., Franke, O.L, and Alley, W.M., 1998, Ground water and surface water—A single resource: U.S. Geological Survey Circular 1139, 79 p.

Wood, W.W., and Fernandez, L.A., 1988, Volcanic rocks, Chapter 39 *in* Back, W., Rosenshein, J.S., and Seaber, P.R., eds., Hydrogeology, v. O-2—The Geology of North America: Boulder, Col., Geological Society of America, p. 353-365.

Appendix A. Assignment of Groundwater Pumpage to Model Layers

In order to allocate the pumpage to model layers, the open intervals for most wells with driller's logs were compiled for all categories except exempt wells and most Washington State Department of Health (WADOH) Group B systems (small public water supply systems with less than 15 connections that also are exempt wells). Open intervals were determined for more than 90 percent of the wells with water rights, and these wells accounted for more than 98 percent of the estimated pumpage associated with water rights. To simplify the large quantity of information for the numerous wells, multiple open intervals were assigned to a well if the interval was greater than 200 ft. For example, if a well was open over 1,000-ft interval, five intervals would be assigned to the well. Estimated pumpage for each well was distributed equally among the intervals, and withdrawals from each interval were considered a separate well. Therefore, the total number of wells in the model is greater than the number of actual wells. During the long, 42-year simulation period, some wells were either deepened or abandoned and a new well was drilled. To the extent possible, these changes are accounted for in the model. For example in the case of a well deepening, pumpage may have been eliminated in the model in one layer, and new pumpage was included in the model in a lower layer. For the case of a new well, if it was not known if an existing well was abandoned, both wells were assumed to pump and the estimated withdrawals were allocated equally between the wells.

For the case of exempt wells, claims and most Group B system wells, several methods were used to allocate estimated pumpage to a model layer. The exempt well pumpage (already distributed by 3,718 census blocks by Sumioka and Welch [U.S. Geological Survey, written commun., 2008]) was assigned to one well in the centroid of the block and the well depth (withdrawal depth) was assigned the median depth value of all non-irrigation/municipal wells in that census block. The median well depth for the exempt wells was 150 ft, which is the same as the reported median for 21,000 wells in the basin (Vaccaro and Sumioka, 2006), and is consistent with the distribution of well depths in the basin (fig. 4). Groundwater claims were represented with 915 wells, and if a well log could not be identified for a claim, the depth was estimated as the median for all wells in the section that the well was in (resulting in a median depth of 140 ft and ranged in depth from 45 to 543 ft) and the location was the center of the section. About 97 percent of the claims withdraw water from the basin-fill units, which is consistent with the pre-1946 development in the basin occurring in the structural basins. For Group B systems for which a well log could not be identified, the well location was based on WADOH information and the depth was estimated as described for the exempt wells; there were a total of 1,022 wells included in the model for these systems. For the case of any other well for which a well log could not be identified, the location was based on the Washington State Department of Ecology information and the depth estimated as described above.

Appendix B. Estimates of Pumpage for Pending Groundwater Applications

To estimate pumpage for the 891 pending groundwater applications several steps were completed. The Washington State Department of Ecology (WADOE) provided a file of the pending applications with a preliminary estimate of the mapped hydrogeologic unit (aquifer designation) that a well for an application would be completed in (J. Kirk, Washington State Department of Ecology, written commun., 2009). There were 450 designations for basalt units and 441 designations for basin-fill sediments. Five of the applications were for frost protection and no quantity was estimated for these applications because the potential use of frost protection pumpage in the future is unknown. As a result, pumpage was estimated for 886 applications (444 basalt units and 442 basin-fill sediments). Spatially, 10 percent of the applications were in Kittitas County, 31 percent in Benton County, and 59 percent in Yakima County. For wells finished in the basin-fill sediments, and the Saddle Mountains and Wanapum units, the opening of the well was assigned to the center of the unit, and for Grande Ronde unit wells, the well was assigned to the Grande Ronde interflow zone nearest a depth of 500-ft below the top of the Grande Ronde unit.

WADOE has developed annual quantity assumptions for applications (J. Kirk, Washington State Department of Ecology, written commun., 2008), for example, 'use 4 af/acre' if "acres irrigated" is listed'. However, it is known that the actual quantity pumped usually is less than the appropriated quantity (Collins, 1987; Cline and Knadle, 1990; Cline and Collins, 1992, 1993; Vaccaro and Sumioka, 2006), and the goal was to estimate a likely actual pumpage estimate from an application that would be consistent with the estimates for the other groundwater withdrawals. Therefore, methods were developed to generate application pumpage estimates consistent with those of Vaccaro and Sumioka (2006). All applications that had an associated irrigated area were initially assigned a value of 3 acre-ft, which is similar to the average value estimated by Vaccaro and Sumioka (2006) for the irrigation wells in the basin. Next, a regression relation

between the maximum gallon per minute (allowable usage for the existing water rights and the estimated pumpage of Vaccaro and Sumioka (2006) was developed. For applications for other uses and without any irrigated area, this regression equation was used to estimate pumpage for an application. Some applications had irrigated acreage but the primary use was not for irrigation (for example, public water supply), that is, there was a reasonable amount of water requested in terms of maximum gallons per minute but with a small irrigated area. For these applications, the regression relation was used. The applications for a single domestic use were assigned a value of 1 acre-ft/yr, regardless of the amount of requested irrigated acreage and gallons per minute because 1 acre-ft is a quantity that WADOE historically has approved for a single domestic use. For the case of applications listing multiple uses (for example, irrigation, domestic multiple, and dairy), the requested gallons per minute and the irrigated acreage were compared in order to estimate which method (regression or water-duty per acre) was to be used. For example, if the irrigated acreage was small and the gallons per minute was large, then the regression equation would be used.

Future pumpage based on a WADOE assignment of an annual allowable rate may vary from the requested amount in the application, especially because the requested amount as a rate (gallons per minute) would need to be related to an allowable annual quantity in acre-feet per year. In some cases the assigned annual rate may be much lower than requested; in other cases, the assigned rate may be the same as the requested amount, and in other cases, an application may be denied. Numerous factors would affect these decisions in the future and thus, it is not possible to make direct estimates of what the actual pumpage from pending applications would be. However, the estimates that were derived are based on historical information and thus, are consistent with the existing-conditions pumpage. The above is especially true because about 78 percent of the application pumpage is for irrigation of more than 60,000 acres.